ERGEBNISSE DER MATHEMATIK
UND IHRER GRENZGEBIETE
HERAUSGEGEBEN VON DER SCHRIFTLEITUNG
DES
„ZENTRALBLATT FÜR MATHEMATIK“
DRITTER BAND
2

THEORY OF LINEAR CONNECTIONS

BY

D. J. STRUIK

BERLIN
VERLAG VON JULIUS SPRINGER
1934

ISBN 978-3-642-50490-7 ISBN 978-3-642-50799-1 (eBook)
DOI 10.1007/978-3-642-50799-1

SOFTCOVER REPRINT OF THE HARDCOVER 1ST EDITION 1934

DEDICATED TO

PROFESSOR J. A. SCHOUTEN

A LEADER IN THE FIELD

OF THE NEW DIFFERENTIAL GEOMETRY

Preface.

This monograph intends to give a general survey of the different branches of the geometry of linear displacements which so far have received attention. The material on this new type of differential geometry has grown so rapidly in recent years that it is impossible, not only to be complete, but even to do justice to the work of the different authors, so that a selection had to be made. We hope, however, that enough territory is covered to enable the reader to understand the present state of the theory in the essential points.

The author wishes to thank several mathematicians who have helped him with remarks and suggestions; especially Dr. J. A. SCHOUTEN of Delft and Dr. N. HANSEN BALL of Princeton.

Cambridge, Mass., October 1933.

D. J. STRUIK.

Contents.

Introduction.

The theory of linear displacement is a result of an investigation into the foundations of differential geometry and also into the structure of geometry as a whole. Though based upon the analysis of space conception undertaken by RIEMANN in 1854[1], it received its impetus only with the advent of general relativity in 1916[2]. Here, space-time is interpreted as a RIEMANNian manifold which is locally euclidean of the MINKOWSKI type, so that the question arises how the comparison between the euclidean world at different points is being performed. This led to the discovery of parallelism in a RIEMANNian manifold[3], then to the extension of this parallelism to manifolds of a more general type. The essential character of the local space changed, in these investigations, from euclidean to affine, the character of the general manifold from RIEMANNian to what is now called affine with or without torsion. This is the principal idea of the work done from 1917 to 1924 by WEYL, SCHOUTEN and EDDINGTON[4]. Closely connected with these investigations are the basic papers of HESSENBERG and KÖNIG[5], which deal with the purely mathematical aspect of the space problem only. From the beginning there has always been an intimate relation between the various attempts to improve or to generalize the theory of relativity and the systematical development of the mathematical theory. For evidence there is, for instance, the recent monograph of VEBLEN[6].

The development of the theory proceeded mainly in three directions. In the first place there came the further study of the displacement of a vector in the tangent space of the manifold (X_n), the generalization of parallel transportation in the sense of LEVI-CIVITA. Representative of this stage in the theory is SCHOUTEN's book "Der RICCI-Kalkül" (1924)[7]. In this sphere also falls a displacement which received considerable attention when EINSTEIN proposed it as a possible space-time manifold[8], and the related theory of HERMITian displacements[9].

A second mode of attack focusses not so much on the displacement of a vector as on the lines of constant direction of the connection, the

[1] RIEMANN: 1854 (1). [2] EINSTEIN: 1916 (2).

[3] LEVI-CIVITA, 1917 (1); SCHOUTEN, 1918 (1).

[4] WEYL: 1918 (2) — 1918 (3) — 1923 (8). — SCHOUTEN: 1924 (5). — EDDINGTON: 1921 (1) — 1923 (9).

[5] HESSENBERG: 1916 (1). — KÖNIG: 1919 (1) — also 1920 (1) — and 1932 (5).

[6] VEBLEN: 1933 (1). [7] SCHOUTEN: 1924 (5).

[8] EINSTEIN: 1928 (2). — See E. BORTOLOTTI: 1929 (8).

[9] SCHOUTEN and VAN DANTZIG: 1930 (6).

so-called "paths". In this case the starting point is a system of ∞^{2n-2} curves in an n-dimensional manifold, which can be defined by a system of ordinary differential equations of the second order. It is natural to inquire for the different kinds of connection compatible with the system of curves as paths. This leads to projective transformations of a displacement and to projective invariants. A similarity between these transformations and the conformal transformations of a RIEMANNian manifold leads to conformal invariance[1]. The field was opened in 1922 by VEBLEN and EISENHART, its method underlies especially the work of VEBLEN and T. Y. THOMAS and EISENHART's "Non-Riemannian Geometry" (1927)[2].

A third theory seems, however, to embrace all the others. It is connected with the work of E. CARTAN who established it and has been developing it since 1922; it appeared for the first time in a paper by KÖNIG[3]. This theory substitutes for the displacement of a vector as primary element the mapping of a space at a point of a manifold on a space at a point in the infinitesimal neighborhood. Displacement of a vector in the affine connections causes such a mapping, but a special variety, namely the affine mapping of affine spaces. It is, however, just as possible to map local spaces projectively upon each other, or conformally. The local space does not need even to be the tangent space; it may differ from it in fundamental group and in number of dimensions. The displacement is then not necessarily a vector displacement; it may be a point displacement, a sphere displacement, a displacement of a line complex, etc. Differential geometry in this stage becomes the study of an n-dimensional manifold X_n, with each point P of which is associated a space S_k defined by a transformation group and of k dimensions, and such that the spaces S_k are related by a law defining the comparison of the S_k at P with the S_k at a point P' of the X_n at infinitesimal distance[4].

It could now be shown that the projective and conformal theory need not be derived from the affine or RIEMANNian theory, but that they are capable of independent foundation. Just as either the classical affine or the classical projective geometry can be taken as the primary element, and the other derived from it, so can the "curved" affine and projective geometry; the same may hold for the conformal, the euclidean and the projective geometry, though this has not yet been satisfactorily shown. To projective geometry a fair amount of study has been devoted, so that the independent structure of this con-

[1] This point is already in WEYL: 1921 (2). — VEBLEN: 1922 (5). — EISENHART: 1927 (1). — See VEBLEN: 1933 (1).

[2] EISENHART-VEBLEN: 1922 (3). — VEBLEN: 1922 (4).

[3] CARTAN: 1922 (6) — 1923 (1), (2) — 1924 (1), (2). — KÖNIG: 1919 (1).

[4] SCHOUTEN: 1926 (1).

nection is well established[1]. Recent attempts of EINSTEIN and others to establish a more comprehensive theory of relativity can also be interpreted in the frame of this geometry[2]. Even the DIRAC theory of the spinning electron has turned out to be an analysis of HERMITIAN quantities fitting into the generalized differential geometry[3].

It is clear that this type of geometry seems far removed from the principles laid down in KLEIN's program of Erlangen. In fact, this program, despite its tremendous influence on the geometrical thought of the last sixty years, was already in a certain respect antiquated at the moment it was conceived. RIEMANN's conceptions on general manifolds went beyond the scope of the Erlangen program. In the infinitesimal neighborhood of a point, however, KLEIN's conceptions hold, even in RIEMANNian manifolds and in the other manifolds of the theory of linear connections. The new theory therefore does not break with KLEIN's program, but generalizes it and gives it a new content[4].

There are several directions in which this theory of linear displacements has again been generalized. A series of papers have discussed the case for which the displacement does not only depend on the points of the X_n, but on the line elements. This work dates back to FINSLER and BERWALD; for a recent exposition we may refer to KAWAGUCHI[5]. Another method is to let the displacement be a displacement dependent on the points of the X_n, but to introduce mapping of line elements of the local spaces. This has been suggested by WIRTINGER. We may even combine the first and the second methods of generalization[6]. Linear displacements may be defined in function-space[7]. And finally, we may give up the linearity of the displacement, which leads to connections, some of which have already been studied by PASCAL[8].

The mathematics to be used in these theories is the so-called tensor analysis, or calculus of RICCI[9]. In the course of years it has undergone considerable change but the central idea of this method has been preserved. In this monograph we shall use the notation and terminology suggested by and under the influence of SCHOUTEN[10], a notation which

1 VAN DANTZIG: 1932 (1) — 1932 (2).

2 See VEBLEN: 1933 (1). — SCHOUTEN and VAN DANTZIG: 1932 (4). — SCHOUTEN: 1933 (2).

3 SCHOUTEN: 1931 (18).

4 CARTAN: 1924 (3). — SCHOUTEN: 1926 (1). — VEBLEN-WHITEHEAD: 1932 (17) p. 31.

5 FINSLER: 1918 (4). — KAWAGUCHI: 1932 (12).

6 WIRTINGER: 1922 (2). — KAWAGUCHI: 1931 (14).

7 KAWAGUCHI: 1929 (15). — MICHAL: 1928 (13). — Comp. MICHAL-PETERSON: 1931 (13).

8 PASCAL: 1903 (1). — See also NOETHER: 1918 (5).

9 RICCI: 1884 (1) and later.

10 See VAN DANTZIG: 1932 (1), (2). — GOŁAB: 1930 (13). — SCHOUTEN: 1924 (5).

allows us to deal with all cases in a uniform way, and to preserve at all times, throughout the fog of the computational work, the guiding geometrical principles.

Textbooks illustrating the development in different stages are the books of WEYL, SCHOUTEN, VEBLEN, EISENHART[1]. There are also several papers which give comprehensive accounts. We mention those of SCHOUTEN, VEBLEN, CARTAN, STRUIK, BORTOLOTTI, WEATHERBURN, EISENHART[2]. Extensive bibliographies of the subject as a whole or of parts of it are found in the textbooks mentioned and also in papers by STRUIK, HLAVATÝ, VAN DANTZIG, and others[3].

[1] WEYL: 1918 (2) — 1923 (8). — SCHOUTEN: 1924 (5). — VEBLEN: 1927 (2) — 1933 (1). — EISENHART: 1927 (1).

[2] SCHOUTEN: 1923 (5) — 1926 (1). — VEBLEN: 1923 (7). — BORTOLOTTI: 1929 (8) — 1931 (3). — CARTAN: 1924 (3) — 1925 (1). — STRUIK: 1925 (4) — 1927 (3). — EISENHART: 1933 (7). — WEATHERBURN: 1933 (8).

[3] STRUIK: 1927 (3). — HLAVATÝ: 1932 (7). — VAN DANTZIG: 1932 (1).

Chapter I.
Algebra.

1. Vectors and tensors in E_n. The starting point in the investigation is the geometry of an affine space of n dimensions E_n and the corresponding tensor algebra. Such a space can be defined as an ordinary euclidean space of n dimensions R_n, in which only those properties which are invariant under the group of affine transformations are studied. For our purpose we confine ourselves to the subgroup which leaves the origin invariant. The transformations of this group, $\mathfrak{A}_n$, can be given by the equations

$$x^{\varkappa'} = \sum_{\varkappa} A_{\varkappa}^{\varkappa'} x^{\varkappa} = A_{\varkappa}^{\varkappa'} x^{\varkappa} ,$$

$$\Delta = \left| A_{\varkappa}^{\varkappa'} \right| = \text{Determinant of the } A_{\varkappa}^{\varkappa'} \neq 0 \qquad \begin{array}{l} \varkappa, \lambda, \mu, \nu, = \cdots = 1, 2, \cdots, n \\ \varkappa', \lambda', \mu', \nu', = \cdots = 1', 2', \cdots n' \end{array}$$

where the $x^{\varkappa}$, $x^{\varkappa'}$ represent the oblique CARTESian coordinates of a point before and after the transformation in the coordinate systems that we can indicate by $(\varkappa)$ and $(\varkappa')$; the $A_{\varkappa}^{\varkappa'}$ are constants. The sign Σ is omitted in accordance with the usual convention. The inverse transformations can be given by

$$x^{\varkappa} = \sum_{\varkappa'} A_{\varkappa'}^{\varkappa} x^{\varkappa'} = A_{\varkappa'}^{\varkappa} x^{\varkappa'} .$$

In such an E_n we can define contravariant, covariant and mixed tensors[1] in the ordinary way. The notation can be seen from this example:

$$v^{\varkappa_1' \dots \varkappa_q'}_{\dots\dots\lambda_1' \dots \lambda_r'} = A^{\varkappa_1' \dots \varkappa_q' \lambda_1 \dots \lambda_r}_{\varkappa_1 \dots \varkappa_q \lambda_1' \dots \lambda_r'} v^{\varkappa_1 \dots \varkappa_q}_{\dots\dots\lambda_1 \dots \lambda_r} .$$

This is a transformation of a mixed tensor of order $q + r$, of contravariant order q, and covariant order r, and

$$A^{\varkappa_1' \dots \varkappa_q' \lambda_1 \dots \lambda_r}_{\varkappa_1 \dots \varkappa_q \lambda_1' \dots \lambda_r'} = A_{\varkappa_1}^{\varkappa_1'} A_{\varkappa_2}^{\varkappa_2'} \dots A_{\varkappa_q}^{\varkappa_q'} A_{\lambda_1'}^{\lambda_1} A_{\lambda_2'}^{\lambda_2} \dots A_{\lambda_r'}^{\lambda_r} .$$

The effect of a coordinate transformation is therefore to change the indices but to leave the central letter (in our case v) unchanged. This central letter stands for the geometrical entity represented by the tensor, an arrow, a plane, a transformation, a complex, etc. The central principle of vector analysis, and of all direct notation, namely the computa-

[1] Following the general use, we speak of *tensors*. Often the word *affinor* is used for what we call tensor; the word *tensor* is then used for what we call a *symmetrical tensor*. The term *polyadic* (dyadic, etc.) has become obsolete. Instead of the term *order* the term *valence* has been recently used.

tion with the geometrical entities themselves, is in this way carried into tensor calculus. The connection of two or more entities by multiplication, done in direct notation by special symbols, is here performed by agreements about the indices.

Two special symbols however are further required, a symbol for symmetrical multiplication and a symbol for alternating multiplication, e. g.

$$v_{(\lambda} w_{\mu\nu)} = \frac{1}{3!}(v_\lambda w_{\mu\nu} + v_\mu w_{\nu\lambda} + v_\nu w_{\lambda\mu} + v_\mu w_{\lambda\nu} + v_\nu w_{\mu\lambda} + v_\lambda w_{\nu\mu})$$

$$v_{[\lambda} w_{\mu\nu]} = \frac{1}{3!}(v_\lambda w_{\mu\nu} + v_\mu w_{\nu\lambda} + v_\nu w_{\lambda\mu} - v_\mu w_{\lambda\nu} - v_\nu w_{\mu\lambda} - v_\lambda w_{\nu\mu}).$$

We also use these brackets to denote the symmetrical or alternating part of a tensor, e. g.

$$v^{(\varkappa\lambda)} = \frac{1}{2!}(v^{\varkappa\lambda} + v^{\lambda\varkappa})$$

$$w_{[\lambda_1\lambda_2\lambda_3\lambda_4]} = \frac{1}{4!}(w_{\lambda_1\lambda_2\lambda_3\lambda_4} - w_{\lambda_2\lambda_1\lambda_3\lambda_4} + w_{\lambda_2\lambda_3\lambda_1\lambda_4} - \text{etc., in total 24 terms}).$$

Such an alternating tensor $v_{[\lambda_1\lambda_2\ldots\lambda_q]}$ is called a *q-vector*. There is also a mixed tensor $A_\lambda^\varkappa$ with components 1 (if $\varkappa = \lambda$) and 0 (if $\varkappa \neq \lambda$) in all coordinate systems. This is the *unit tensor* and it follows equations such as $v_{\lambda\varkappa} A_\mu^\varkappa = v_{\lambda\mu}$. It should not be confused with the so-called KRONECKER symbol $\delta_\lambda^\varkappa$, which is simply a matrix of n^2 numbers equal to 1 when $\varkappa = \lambda$ and to 0 when $\varkappa \neq \lambda$. The $\delta_\lambda^\varkappa$ have nothing to do with transformations. The unit tensor is therefore a mixed tensor, the components if which in every coordinate system are given by the KRONECKER symbol.

2. Densities. The volume of an n-dimensional volume in E_n is an invariant under the group $\mathfrak{A}_n$ only when $\Delta = 1$. When $\Delta \neq 1$ a transformation multiplies the value by Δ. We call a quantity $\mathfrak{p}$ which behaves in that way a *scalar density* of weight -1. Densities are written with a Gothic letter. A scalar density of weight $+\mathfrak{k}$ is defined by its transformation

$$\overset{(\varkappa')}{\mathfrak{p}} = \Delta^{-\mathfrak{k}} \overset{(\varkappa)}{\mathfrak{p}} \quad \text{(the } \varkappa,\ \varkappa' \text{ indicating the coordinate systems).}$$

A tensor density of weight $+\mathfrak{k}$ is defined by the transformation

$$\mathfrak{v}^{\varkappa_1'\ldots\varkappa_q'}_{\cdots\cdots\lambda_1'\ldots\lambda_r'} = \Delta^{-\mathfrak{k}} A^{\varkappa_1'\ldots\varkappa_q'\lambda_1\ldots\lambda_r}_{\varkappa_1\ldots\varkappa_q\lambda_1'\ldots\lambda_r'} \mathfrak{v}^{\varkappa_1\ldots\varkappa_q}_{\cdots\cdots\lambda_1\ldots\lambda_r}.$$

A contravariant n-vector $v^{\lambda_1\ldots\lambda_n} = v^{[\lambda_1\ldots\lambda_n]}$ has all its components zero except those for which the indices are all different; they are all equal to $v^{12\ldots n}$ or its negative. This component $v^{12\ldots n}$ is itself a scalar density of weight -1 as

$$v^{1'2'\ldots n'} = A^{1'2'\ldots n'}_{\lambda_1\lambda_2\ldots\lambda_n} v^{\lambda_1\lambda_2\ldots\lambda_n} = \frac{1}{n!} A^{[1'2'\ldots n']}_{[1\,2\,\ldots\,n\,]} v^{12\ldots n} = \Delta v^{12\ldots n}.$$

To every scalar density of weight -1 belongs a volume in E_n with an n-dimensional screw-sense, determining a contravariant n-vector, and similarly a covariant n-vector when the weight is $+1$.

An example of a tensor density of weight $+2$ is $|h_{\lambda\mu}| h_{\varkappa\nu}$, where $|h_{\lambda\mu}|$ is the determinant of the tensor $h_{\lambda\mu}$ of rank n. Tensor densities of weight $\mathfrak{k}$ are also called *relative tensors* of weight $-\mathfrak{k}$.[1]

3. Measuring vectors. To every coordinate system $(\varkappa)$ belongs a set of n contravariant measuring vectors $\underset{1}{e}^{\varkappa}, \underset{2}{e}^{\varkappa}, \ldots, \underset{n}{e}^{\varkappa}$, in short $\underset{\lambda}{e}^{\varkappa}$, where $\underset{1}{e}^{\varkappa}$ has components $(1, 0, \ldots, 0)$, $\underset{2}{e}^{\varkappa}$ $(0, 1, 0, \ldots, 0)$, etc., in short

$$\underset{\lambda}{e}^{\varkappa} \overset{*}{=} \delta^{\varkappa}_{\lambda}.$$

The star above the $=$ sign meaning that the equation holds only for a special coordinate system.[2] The components of the $\underset{\lambda}{e}^{\varkappa}$ change when we pass to another coordinate system $(\varkappa')$: $\underset{\lambda}{e}^{\varkappa'} = A^{\varkappa'}_{\varkappa} \underset{\lambda}{e}^{\varkappa}$. In the same way we have n covariant measuring vectors $\overset{1}{e}_{\lambda}, \overset{2}{e}_{\lambda}, \ldots, \overset{n}{e}_{\lambda}$, in short $\overset{\varkappa}{e}_{\lambda}$, satisfying

$$\overset{\varkappa}{e}_{\lambda} \overset{*}{=} \delta^{\varkappa}_{\lambda}.$$

The contravariant measuring vectors determine the edges of an n-dimensional parallelepiped. Its $(n-1)$-dimensional faces can be taken as the covariant measuring vectors, a covariant vector being geometrically represented by an E_{n-1} in the same way as a contravariant vector is represented by a point E_0, in conjunction with the origin. Point $v^{\varkappa}$ and $E_{n-1}\, w_{\lambda}$ are incident if $v^{\lambda} w_{\lambda} = 0$. Covariant measuring vectors selected in this way can therefore be related to the contravariant measuring vectors by the equations

$$\underset{\varkappa}{e}^{\lambda} \overset{\nu}{e}_{\lambda} \overset{*}{=} \delta^{\nu}_{\varkappa}.$$

We have, besides, as a result:

$$\overset{\lambda}{e}_{\varkappa} \underset{\lambda}{e}^{\mu} = A^{\mu}_{\varkappa}.$$

The four symbols $A^{\varkappa}_{\lambda}$, $\delta^{\varkappa}_{\lambda}$, $\overset{\varkappa}{e}_{\lambda}$, $\underset{\lambda}{e}^{\varkappa}$ therefore represent all the same numbers in a fixed coordinate system, but follow different laws of transformation, i. e.

$$A^{\varkappa}_{\lambda} \overset{*}{=} \delta^{\varkappa}_{\lambda} \overset{*}{=} \overset{\varkappa}{e}_{\lambda} \overset{*}{=} \underset{\lambda}{e}^{\varkappa}.$$

From the equations following from the definition

$$A^{\varkappa'}_{\varkappa} A^{\varkappa}_{\lambda'} = A^{\varkappa'}_{\lambda'}$$

---

[1] WEYL: 1918 (3). — VEBLEN-THOMAS: 1924 (8). — THOMAS: 1925 (6). — THOMAS-MICHAL: 1927 (5). — HLAVATÝ: 1928 (10). — SCHOUTEN-HLAVATÝ: 1929 (2).

[2] See a more general application in SCHOUTEN-VAN DANTZIG 1933 (6).

we see that $A_{\lambda'}^{\varkappa'}$ can be taken as the $(\varkappa')$ component of the transformation matrix $A_{\lambda'}^{\varkappa}$. This justifies the use of the same central letter A for transformation matrix and unit tensor.

Tensors can be decomposed with respect to these measuring vectors, e. g.

$$h_{\lambda\mu} \stackrel{*}{=} \underset{\sigma\tau}{h} \overset{\sigma}{e}_{\lambda} \overset{\tau}{e}_{\mu}$$

the $\underset{\sigma\tau}{h}$ are scalars, defined only with respect to the coordinate system of the $\overset{\varkappa}{e}_{\lambda}$.[1] Densities can also be decomposed with respect to these measuring vectors:

$$\mathfrak{v}^{\varkappa} \stackrel{*}{=} \overset{\tau}{\mathfrak{v}} \underset{\tau}{e}^{\varkappa}$$

the $\overset{\tau}{\mathfrak{v}}$ are scalar densities of the same weight as the vector density $\mathfrak{v}^{\varkappa}$.[2]

4. Point algebra. Tensors are defined with respect to a certain group of transformations. On the geometrical interpretation of this group depends the geometrical interpretation of the tensors. It is therefore possible to introduce a projective interpretation, a conformal interpretation, etc. We shall illustrate this by sketching a *point algebra.*

The starting point is an n-dimensional projective space D_n, in which a coordinate $(n+1)$-cell is given by an origin-point P and n linearly independent other basic points. We can now build up a system of homogeneous coordinates, in which P is given by a set $\underset{0}{\mathfrak{u}}^a$, $a = 0, 1, 2, \ldots, n$, and the other basic points by $\underset{i}{\mathfrak{u}}^a$, $i = 1, 2, \ldots, n$. Every other point $\mathfrak{v}^a$ of D_n can be expressed as a linear combination of the $\mathfrak{u}^a$. This brings us to an algebra identical with the point calculus of MÖBIUS[3]. If we consider as essential the components and not their ratio, we have to attach to every point a weight, and consequently we will say that every vector $\mathfrak{v}^a$ represents a *point* of certain degree. We can represent the transformation of points under a change of coordinate system in the following way:

$$\mathfrak{v}^{a'} = \mathfrak{A}_a^{a'} \mathfrak{v}^a, \qquad \begin{array}{l} a, b, \ldots = 0, 1, \ldots, n \\ a', b', \ldots = 0', 1', \ldots, n'. \end{array}$$

Covariant points can be interpreted as D_{n-1} in D_n. They transform in this way:

$$\mathfrak{w}_{b'} = \mathfrak{A}_{b'}^{b} \mathfrak{w}_b .$$

We may, without loss of generality, take the determinant $|\mathfrak{A}_a^{a'}| = 1$.

[1] About this process of "Abdrosselung" see SCHOUTEN-VAN KAMPEN: 1930 (21).

[2] KÖNIG: 1920 (1) — 1932 (5). — SCHOUTEN: 1924 (5).

[3] MÖBIUS: 1827 (1). — See R. MEHMKE: 1913 (1).

We can define in a similar way covariant, contravariant and mixed tensors of higher order and given degree. There is a unit tensor $\mathfrak{A}_b^a$

$$\mathfrak{v}^a = \mathfrak{A}_b^a \mathfrak{v}^b .$$ [1]

We normalize this tensor in such a way that

$$\mathfrak{A}_b^a \overset{*}{=} \delta_b^a .$$

Extension of the vector symbolism to the conformal group has also been investigated[2].

5. The general manifold X_n. Classical differential geometry is obtained by taking a euclidean space R_N (usually, $N = 3$) and imbedding into this R_N certain surfaces V_n (usually $n = 2$). The geometrical properties of the R_N *induce* into the V_n a differential geometry, that is a way to compare the geometrical properties at one point of the V_n with those at a point of the V_n in the immediate neighborhood. The theory of displacements begins differently. It starts with an n-dimensional manifold X_n in the sense of analysis situs, and then sets up a group of postulates by which it is possible to define a differential geometry without the necessity of imbedding the X_n into a metrical manifold of more dimensions. To allow this, the X_n must first satisfy certain general conditions[3], which will allow us to build up a one to one correspondence between a set of points P of this X_n and a set of ordered sets of n real numbers $\xi^\varkappa$, $\varkappa = 1, 2, \ldots, n$ which form a coordinate system $(\varkappa)$ in X_n. The $\xi^\varkappa$ are called the *original variables*. It must be possible to define the coordinate transformations

$$(\mathfrak{G}_n) \qquad \xi^{\varkappa'} = \overset{\varkappa'}{f}(\xi^\varkappa),$$

$$\Delta = \text{Determinant} \left| \frac{\partial \xi^{\varkappa'}}{\partial \xi^\varkappa} \right| \neq 0 \qquad \begin{array}{l} \varkappa, \lambda, \mu, \nu, \ldots = 1, 2, \ldots, n \\ \varkappa', \lambda', \mu', \nu', \ldots = 1', 2', \ldots, n' \end{array}$$

in this X_n in such a way that there is about each point $\overset{}{\underset{0}{\xi}}{}^\varkappa$ a region in which the transformation of the differentials

$$(\mathfrak{G}_n') \qquad d\,\xi^{\varkappa'} = \frac{\partial \xi^{\varkappa'}}{\partial \xi^\varkappa} d\,\xi^\varkappa$$

defines an affine transformation in an E_n. Under circumstances it may also be required that higher derivatives of the functions involved exist. A manifold X_n, in which a differential geometry can be constructed may be called a *regular manifold*, and when we write X_n we always mean such a manifold.

---

[1] Cartan: 1924 (3). — Gołab: 1930 (13).

[2] Cartan: 1923 (2). — See also Blaschke: Differentialgeometrie III.

[3] Veblen-Whitehead: 1932 (17) — 1931 (1). — cfr. also Järnefelt: 1928 (14). — Veblen: 1925 (3).

Transformations $(\mathfrak{G}'_n)$ form a group, which allows us to define in the E_n at a point P of the X_n ("the local E_n") all the tensors and tensor densities defined in the preceding articles. As this is possible at all points of X_n, we are able to define *fields* of tensors and tensor densities, defined as functions of $\xi^\varkappa$, transforming under $(\mathfrak{G}_n)$ in the ordinary way, if we take $A^{\varkappa'}_\varkappa = \partial \xi^{\varkappa'}/\partial \xi^\varkappa$, $A^\varkappa_{\varkappa'} = \partial \xi^\varkappa/\partial \xi^{\varkappa'}$, e. g.

$$v^{\varkappa'} = A^{\varkappa'}_\varkappa v^\varkappa .$$

When, in an X_n, we deal with a field of vectors, tensors, etc., we will simply say that we are dealing with "a vector", "a tensor", etc. if this can be done without ambiguity. We assume analyticity for these functions, though the existence of a certain number of derivatives is sufficient for many purposes.

There are fields of functions of $\xi^\varkappa$ which also are transformed under a coordinate transformation, but not like tensors or tensor densities. A simple example is $\partial v^\varkappa/\partial \xi^\mu$, which transforms as follows

$$\partial_{\lambda'} v^{\varkappa'} = \frac{\partial v^{\varkappa'}}{\partial \xi^{\lambda'}} = A^{\varkappa'\lambda}_{\varkappa\lambda'} \frac{\partial v^\varkappa}{\partial \xi^\lambda} + v^\varkappa \frac{\partial}{\partial \xi^{\lambda'}} A^{\varkappa'}_\varkappa = A^{\varkappa'\lambda}_{\varkappa\lambda'} \partial_\lambda v^\varkappa + v^\varkappa A^\lambda_{\lambda'} \partial_\lambda A^{\varkappa'}_\varkappa ,$$

where we write $\partial_\lambda = \partial/\partial \xi^\lambda$, $\partial_{\lambda'} = \partial/\partial \xi^{\lambda'}$, etc.

We now introduce with VEBLEN the notion of *geometrical object* (more briefly: *object*)[1]. This is a set of N functions of the $\xi^\varkappa$, given in a coordinate system $(\varkappa)$, which obey a transformation law by which we can compute a unique corresponding set of N functions of $\xi^{\varkappa'}$ in the transformed coordinate system, expressed in original functions, the $A^{\varkappa'}_\varkappa$ and their derivatives. If the transformation is linear homogeneous, with the parameters of the transformation as coefficients in the way indicated in art. 1, we have a tensor (special case: scalar, vector). Densities are also geometrical objects. Tensors and densities are called *quantities*. A more general object is the set of CHRISTOFFEL symbols $\left\{ {\varkappa \atop \mu\lambda} \right\}$ belonging to a symmetrical tensor $g_{\lambda\mu}$ which transforms

$$\left\{ {\varkappa' \atop \mu'\lambda'} \right\} = A^{\mu\lambda\varkappa'}_{\mu'\lambda'\varkappa} \left\{ {\varkappa \atop \mu\lambda} \right\} + A^{\lambda\varkappa'}_{\lambda'\varkappa} \partial_\lambda A^\varkappa_{\mu'} ,$$

where the transformation involves only the N "components" $\left\{ {\varkappa \atop \mu\lambda} \right\}$, the parameters of the transformation and their first derivatives. The system $(v^\varkappa, \partial_\lambda v^\varkappa)$ is also a geometrical object and an example of an "absolute system" of VITALI[2] (but not $\partial_\lambda v^\varkappa$ alone).

6. Non-holonomic measuring vectors. In the local E_n at a point P of X_n, we can again introduce two sets of measuring vectors $\underset{\lambda}{e}^\varkappa$ and $\overset{\varkappa}{e}_\lambda$

[1] See VEBLEN-WHITEHEAD: 1932 (17) p. 46.

[2] VITALI: 1929 (24). — Comp. BORTOLOTTI: 1931 (6).

defined with respect to the variables $\xi^\varkappa$. But we can also introduce these measuring vectors independent of the $\xi^\varkappa$; we shall then write $\underset{i}{e}^\varkappa$, $\overset{k}{e}_\lambda$, $i, j, k = 1, 2, \ldots, n$. Then we have a new coordinate system (k) in the E_n, (not necessarily in the X_n) and we can pass from one set of measuring vectors to another set, leading to a new coordinate system (k'). We have

$$\underset{i}{e}^\varkappa \overset{k}{e}_\varkappa \overset{*}{=} \delta_i^k ; \qquad \underset{i}{e}^\varkappa \overset{i}{e}_\lambda = A_\lambda^\varkappa .$$

The components of $\underset{i}{e}^\varkappa$ and $\overset{k}{e}_\lambda$ with respect to (k) can be indicated by $\underset{i}{e}^k$, $\overset{k}{e}_i$. We have A_i^k as the unit tensor in the new coordinate system:

$$\underset{i}{e}^k \overset{*}{=} \overset{k}{e}_i \overset{*}{=} \delta_i^k \overset{*}{=} A_i^k .$$

The components of a vector $v^\varkappa$, w_λ with respect to these coordinates can be denoted by latin indices,

$$v^k = v^\varkappa \overset{i}{e}_\varkappa \underset{i}{e}^k \overset{*}{=} v^\varkappa \overset{k}{e}_\varkappa ,$$

$$w_i = w_\lambda \underset{k}{e}^\lambda \overset{k}{e}_i \overset{*}{=} w_\lambda \underset{i}{e}^\lambda .$$

When both the $\underset{i}{e}^\varkappa$, $\overset{k}{e}_\lambda$ and the original measuring vectors exist in the same E_n, we can give a meaning to a component like $v_{\lambda j}^{\cdot\cdot k}$, namely,

$$v_{\lambda j}^{\cdot\cdot k} = v_{ij}^{\cdot\cdot k} A_\lambda^i , \quad \text{where} \quad A_\lambda^i = \overset{j}{e}_\lambda \underset{j}{e}^i \overset{*}{=} \overset{i}{e}_\lambda ; \quad \text{also} \quad A_i^\varkappa = \underset{j}{e}^\varkappa \overset{j}{e}_i .$$

We are now able to introduce a system of "local" coordinates into the local E_n, defined by means of a vector $x^\varkappa$ with respect to the measuring vectors,

$$x^k = x^\varkappa \overset{i}{e}_\varkappa \underset{i}{e}^k \overset{*}{=} x^\varkappa \overset{k}{e}_\varkappa$$

which coordinates are independent of the $\xi^\varkappa$. At each point P of the X_n such systems can be established. The $\underset{i}{e}^\varkappa$ then build up in the X_n n congruences of curves, but the $\overset{k}{e}_\lambda$ do not necessarily build up n systems of $\infty^1 X_{n-1}$. This is the case only if

$$\partial_{[\mu} \overset{k}{e}_{\lambda]} = 0 , \quad \text{equivalent to} \quad \partial_{[\mu} A_{\lambda]}^k = 0 .$$

Then the $\overset{k}{e}_\lambda$ are gradient vectors, and there exist n independent scalar fields $\overset{k}{\xi}$ such that $\overset{k}{e}_\lambda = \partial_\lambda \overset{k}{\xi}$. These scalar fields can now be taken as original variables $\xi^\varkappa$ in the X_n. If, however, $\partial_{[\mu} \overset{k}{e}_{\lambda]} \neq 0$ there are no such scalar fields and the expression

$$(d\,\xi)^k = \overset{i}{e}_\varkappa \underset{i}{e}^k \, d\,\xi^\varkappa \overset{*}{=} \overset{k}{e}_\varkappa \, d\,\xi^\varkappa$$

is not an exact differential. Then we say that we have in the E_n a *non-holonomic* system of parameters[1]. In such a system we can introduce the same algebra as in a holonomic system, e. g.,

$$v^\lambda = w^{\lambda}_{\cdot\,\nu\mu}\, u^{\nu\mu} \rightarrow v^l = w^{l}_{\cdot\, n s}\, u^{n s},$$

which is the same geometrical relation referred to different coordinate systems.

An example in RIEMANNian geometry is that of the introduction of an orthogonal ennuple, that is a system of n mutually orthogonal congruences. RICCI has often simplified his equations by referring them to such an orthogonal ennuple, taking unit vectors as measuring vectors. We shall return to this in Ch. II.

7. Pseudotensors. It is often necessary to introduce into the X_n apart from the coordinate transformations

$$\xi^{\varkappa'} = f^{\varkappa'}(\xi^{\varkappa}), \quad \Delta = |\, \partial_\varkappa \xi^{\varkappa'} \,| \neq 0$$

a transformation of an auxiliary coordinate ξ°

$$\xi^{\circ\prime} = \tau\, \xi^{\circ}$$

where τ is a function of the $\xi^{\varkappa}$. This allows us to define a *pseudoscalar* $\mathfrak{p}$ of class $\mathfrak{k}$ which transforms in the manner

$$\overset{(\varkappa')}{\mathfrak{p}} = \tau^{\mathfrak{k}} \overset{(\varkappa)}{\mathfrak{p}}, \quad \text{in short } \mathfrak{p}' = \tau^{\mathfrak{k}}\, \mathfrak{p}$$

where $\tau^{\mathfrak{k}}$ is the $\mathfrak{k}$th power of τ, and *pseudotensors* of class $\mathfrak{k}$, as

$$\mathfrak{v}^{\varkappa'\lambda'} = \tau^{\mathfrak{k}} A^{\varkappa'\lambda'}_{\varkappa\lambda}\, \mathfrak{v}^{\varkappa\lambda}.$$

Two cases are possible, τ being either dependent on the transformation of the $\xi^{\varkappa}$, or independent. A special case of dependence is $\tau = \Delta^{-1}$. In this case we get the densities and for this reason we denote pseudoquantities also with a gothic letter. The other case is new.

To the coordinate ξ° belongs a measuring scalar $\mathfrak{e}$ of class 1 with one component of value 1 for this special coordinate system. When ξ° is transformed to $\xi^{\circ\prime} = \tau \xi^{\circ}$, we have a new measuring scalar $\mathfrak{e}'$, with component 1 in the new system. Hence in the old system

$$\mathfrak{e}' \overset{*}{=} \tau\, \mathfrak{e} \overset{*}{=} \tau.$$

To every pseudotensor of class $\mathfrak{k}$ belongs an ordinary tensor with the same components with respect to $\xi^{\varkappa}$, ξ°, e. g.

$$v_{\mu\lambda}^{\cdot\cdot\varkappa} \overset{*}{=} \mathfrak{e}^{-\mathfrak{k}}\, \mathfrak{v}_{\mu\lambda}^{\cdot\cdot\varkappa}.$$ [2]

Pseudotensors, like tensors, are quantities. They appear often in an intermediate state of the theory, when it is necessary to single out one variable.

[1] VRANCEANU: 1926 (4). — HORAK: 1927 (8). — SCHOUTEN: 1929 (4). — Comp. also HESSENBERG: 1916 (1). — SCHOUTEN: 1918 (1). — CARTAN: 1923 (1). — HLAVATÝ: 1924 (9). — VRANCEANU: 1928 (15).

[2] SCHOUTEN-HLAVATÝ: 1929 (2).

Chapter II.
Affine connections.

1. The principle of displacement. In euclidean geometry it is possible to move a vector parallel to itself from one point to another point at finite distance. This means that in this geometry a law is given by which it is possible to associate in a unique way a vector to every point in space, if a vector is given at one point. The length of one vector and the angle between two vectors are invariant under such a parallel displacement.

This parallelism allows us to compare vectors at different points of euclidean space as to length and direction. By parallel displacement one vector can be brought to the point at which the other vector is, after which comparison can be made by purely local means.

On this principle is based the method of the moving trihedron which plays an important role in the differential geometry of curves and surfaces. In the case of a surface V_2 in euclidean space R_3, we have connected with each point P of V_2 a local trihedron built up by two vectors in the tangent R_2 and the surface normal. It is useful to express this by saying that with every point of V_2 a local R_3 is associated. The moving trihedron method allows us to compare the local R_3 at different points of the V_2.[1] In this case we can combine the local R_3 into one "collective" R_3. We shall see that this is a special case from the point of view of displacement theory.

An entirely different case was presented by LEVI-CIVITA and SCHOUTEN[2]. They showed how it is possible to connect with a RIEMANNian geometry an intrinsic parallelism, which does not require the imbedding of the RIEMANNian manifold V_n in a euclidean space of more than n dimensions. In this displacement parallelism is defined for points at infinitesimal distance in a given direction. The length of the vector and the angle between two vectors again remain invariant. With the aid of this a covariant differential is defined

$$\delta v^{\varkappa} = d v^{\varkappa} + \left\{ {\varkappa \atop \mu\lambda} \right\} v^{\mu} d\xi^{\lambda},$$

where $\delta v^{\varkappa}$ is again a vector. The parallel displacement along a curve is uniquely determined, but not for two points connected by different curves. We can express this kind of displacement by saying that with

---

[1] DARBOUX: 1889 (1) Livre V Ch. I.

[2] LEVI-CIVITA: 1917 (1). — SCHOUTEN: 1918 (1). The method of LEVI-CIVITA still required imbedding, though his result was intrinsic. SCHOUTEN, however, used an intrinsic method.

every point of the V_n a local R_n (the euclidean tangent space, or the local tangent space of the first differentials) is associated, and the laws of this parallelism allow us to compare the local R_n at different points of the V_n.

A third case, seeming entirely separated, is CLIFFORD parallelism in elliptic space. Here we can define, in two different ways, a direction through a point parallel to a given direction through another point at finite distance.

The theory of linear displacements has unified all these points of view. All three cases appear now as specializations of a general theory, in which we associate with every point of an X_n a local space S_k and build up laws to compare these local spaces.

To understand this better, we sketch this point of view for the second case, that of the RIEMANNian connection V_n.

At two points P, P' at infinitesimal distance exist two local tangent R_n, one belonging to P, the other, R'_n, to P'. In each R_n is a system of reference, e. g., a CARTESian coordinate system. An observer at P can think that he is in an R_n; he can, for a given V_n, also localize in this R_n the point P' and the coordinate system of the R'_n at P', which have a definite position with respect to the coordinate system in the R_n at P. If we now consider a series of local R_n along a curve PQ of V_n, then the observer at P will be able to localize successively, in the same R_n, all the different R'_n of the points Q' of the curve PQ. The curve PQ is thus *developed*, with its different corresponding R_n, on the R_n at P. The observer at A will only be aware that he is not in an R_n, but in a manifold of different connection, when he localizes in his space R_n the point Q and its coordinate system, once by developing the V_n along one curve PQ of V_n, and another time along another curve PQ of V_n. It is not a priori obvious that he will get the same point and coordinate system. If we now take for PQ an infinitesimal closed curve, Q falling on P, then it can be proved that in the case of LEVI-CIVITA parallelism the point Q will always come in the same place (we call this absence of *torsion*; see art. 3) but the coordinate system will turn. In this the *curvature* of the V_n reveals itself. Other connections can be constructed by modification of the local space or introduction of torsion[1].

2. Affine displacement L_n. The first generalization of the parallelism of LEVI-CIVITA was obtained by associating with every point of the X_n a local E_n. This is natural, as we can take as local E_n the tangent E_n to the X_n, the existence of which is established by the definition of X_n. The geometry thus obtained is called the geometry of the *affine connection* and we shall denote it by L_n.

---

[1] CARTAN: 1924 (3), (4) — 1925 (1) — 1930 (9).

Such a connection is defined by an *affine displacement*[1]. This can be done by the definition of a covariant differential which allows us to compare the local E_n at a point P with the local E_n at a point P', at infinitesimal distance.

In the X_n we again introduce a group of transformations of the original variables,

(2.1) $$\xi^{\varkappa'} = f^{\varkappa'}(\xi^{\varkappa}), \qquad \Delta = |\partial_{\varkappa}\xi^{\varkappa'}| \neq 0$$

implying

(2.2) $$d\xi^{\varkappa'} = A^{\varkappa'}_{\varkappa} d\xi^{\varkappa}, \qquad A^{\varkappa'}_{\varkappa} = \partial_{\varkappa}\xi^{\varkappa'}.$$

We introduce fields of geometrical objects, in the first place, fields of quantities (i. e. tensors and densities), e. g.:

$$v^{\varkappa'\lambda'\mu'} = A^{\varkappa'\lambda'\mu'}_{\varkappa\lambda\mu} v^{\varkappa\lambda\mu}.$$

For local E_n we will now take the tangent E_n in which these quantities behave, at each point, like ordinary affine quantities (Ch. I), and in which furthermore pseudotensors can be defined.

Then we define the covariant differential δT of a quantity in the following way:

1. Every quantity has a covariant differential depending on this quantity, its first ordinary derivatives, and on a direction of progress $d\xi^{\varkappa}$.

2. The components of a quantity and of its covariant differential transform in the same way under (2.1) and (2.2). This means that the covariant differential of a vector is again a vector, etc.

3. The covariant differential is a linear homogeneous integral function of the $d\xi^{\varkappa}$.

4. Covariant differentiation of a sum or of an outer product of two quantities T and U follows the ordinary formal rules. Hence $\delta(T+U) = \delta T + \delta U$; $\delta(TU) = (\delta T)U + T(\delta U)$.

5. Rule 4. also holds for the inner product. Hence

$$\delta(v^{\varkappa} w_{\varkappa}) = (\delta v^{\varkappa}) w_{\varkappa} + v^{\varkappa}\delta w_{\varkappa},$$ [2]

$$\delta A^{\varkappa}_{\lambda} = 0.$$

From these rules follow other rules.

a) The covariant derivative of a scalar is the ordinary derivative.

b) The covariant derivative of a vector is of the form

$$\delta v^{\varkappa} = dv^{\varkappa} + \Gamma^{\varkappa}_{\mu\lambda} v^{\lambda} d\xi^{\mu},$$
$$\delta w_{\lambda} = dw_{\lambda} - \Gamma^{\varkappa}_{\mu\lambda} w_{\varkappa} d\xi^{\mu},$$

---

[1] Following a suggestion by VEBLEN we use *displacement* (Übertragung) when there is an infinitesimal transportation of quantities. The manifold X_n obtains a certain *connection* (Zusammenhang), when there is also a covariant derivative.

[2] SCHOUTEN-HLAVATÝ: 1929 (2). Omission of 5. leads to different $\Gamma^{\varkappa}_{\mu\lambda}$ for covariant and for contravariant quantities. — See SCHOUTEN: 1924 (5) Ch. II. Introduction of densities makes this discrimination superfluous.

where the $\Gamma^{\varkappa}_{\mu\lambda}$ form a geometrical object with n^3 components, dependent on the $\xi^{\varkappa}$ and not on the vector field[1]. The difference in sign of the second term of the second member is a result of rule 5. For these $\Gamma^{\varkappa}_{\mu\lambda}$ the equations

$$\Gamma^{\varkappa}_{\mu\lambda}\, d\xi^{\mu} \overset{*}{=} \delta \underset{\lambda}{e}^{\varkappa}; \qquad \Gamma^{\varkappa}_{\mu\lambda}\, d\xi^{\mu} \overset{*}{=} \delta \overset{\varkappa}{e}_{\lambda}$$

hold.

c) Covariant derivatives of tensors follow rules similar to those of RIEMANNian geometry, e. g.

$$\delta v^{\varkappa}_{\cdot\lambda\nu} = d v^{\varkappa}_{\cdot\lambda\nu} + \Gamma^{\varkappa}_{\mu\pi} v^{\pi}_{\cdot\lambda\nu}\, d\xi^{\mu} - \Gamma^{\pi}_{\mu\lambda} v^{\varkappa}_{\cdot\pi\nu}\, d\xi^{\mu} - \Gamma^{\pi}_{\mu\nu} v^{\varkappa}_{\cdot\lambda\pi}\, d\xi^{\mu}.$$

d) For the covariant differentials of pseudoscalars of class $\mathfrak{k}$ we find, if we introduce a coordinate transformation $\xi^{0'} = \tau \xi^0$

$$\delta \mathfrak{p} = d\mathfrak{p} - \mathfrak{k}\, \Gamma_{\mu}\, d\xi^{\mu}$$

where Γ_{μ} are a set of n functions of the $\xi^{\varkappa}$ defining a geometrical object. As long as the function τ is independent of the transformation of the $\xi^{\varkappa}$, which defines tensors and densities, the Γ_{λ} are independent of the $\Gamma^{\varkappa}_{\mu\lambda}$. For densities of weight $-\mathfrak{k}$ there exists a relation which can be found from the covariant differential of a covariant n-vector (Ch. I, 2). Such an n-vector defines a density of weight $+1$; hence

$$\delta w_{\lambda_1 \dots \lambda_n} = d w_{\lambda_1 \dots \lambda_n} - \Gamma^{\varkappa}_{\mu\varkappa} w_{\lambda_1 \dots \lambda_n}\, d\xi^{\mu}.$$

From this we derive for densities the relation between Γ_{λ} and $\Gamma^{\varkappa}_{\mu\lambda}$

$$\Gamma_{\lambda} = \Gamma^{\varkappa}_{\lambda\varkappa}.$$

e) For the covariant differentials of pseudotensors of class $\mathfrak{k}$ the formula is therefore as in this example:

$$\delta \mathfrak{U}_{\lambda}^{\cdot\varkappa} = d\mathfrak{U}_{\lambda}^{\cdot\varkappa} + \Gamma^{\varkappa}_{\mu\nu} \mathfrak{U}_{\lambda}^{\cdot\nu}\, d\xi^{\mu} - \Gamma^{\nu}_{\mu\lambda} \mathfrak{U}_{\nu}^{\cdot\varkappa}\, d\xi^{\mu} - \mathfrak{k}\, \mathfrak{U}_{\lambda}^{\cdot\varkappa}\, \Gamma_{\mu}\, d\xi^{\mu}.$$

When the coordinate system is transformed from $(\varkappa)$ to $(\varkappa')$, we have (see Ch. I, art. 6)

$$\Gamma^{\varkappa'}_{\mu'\lambda'} = A^{\varkappa'\mu\lambda}_{\varkappa\mu'\lambda'} \Gamma^{\varkappa}_{\mu\lambda} + A^{\varkappa'}_{\nu}\, \partial_{\mu'} A^{\nu}_{\lambda'} = A^{\varkappa'\mu\lambda}_{\varkappa\mu'\lambda'} \Gamma^{\varkappa}_{\mu\lambda} - A^{\lambda\mu}_{\lambda'\mu'}\, \partial_{\mu} A^{\varkappa'}_{\lambda},$$
$$\Gamma_{\lambda'} = A^{\lambda}_{\lambda'} \Gamma_{\lambda} + \partial_{\lambda'} \ln \tau.$$

The latter formula shows that as long as τ is independent of the parameters of the transformation of the $\xi^{\varkappa}$, the Γ_{λ} behave like the components of a vector when ξ^0 does not vary.

The covariant derivative can be found from the definition:

$$\delta T = d\xi^{\mu}\, \nabla_{\mu} T.$$

Hence

$$\nabla_{\mu} \mathfrak{U}_{\lambda}^{\cdot\varkappa} = \partial_{\mu} \mathfrak{U}_{\lambda}^{\cdot\varkappa} + \Gamma^{\varkappa}_{\mu\nu} \mathfrak{U}_{\lambda}^{\cdot\nu} - \Gamma^{\nu}_{\mu\lambda} \mathfrak{U}_{\nu}^{\cdot\varkappa} - \mathfrak{k}\, \mathfrak{U}_{\lambda}^{\cdot\varkappa} \Gamma_{\mu}.$$

This symbol ∇ is taken from ordinary vector analysis, it is HAMILTON's "nabla" operator. It behaves algebraically like a covariant

---

[1] DOUGLAS: 1928 (1) writes $-\Gamma$ for our Γ.

vector, and enters therefore into tensor calculus with one covariant index.

A displacement for which the covariant differential is zero will be called a *parallel* displacement. It can be uniquely defined along the arc of a curve in X_n. In this way the local E_n at a point P of X_n is mapped with its body of vectors on the local E_n along the curve. This mapping is also an affine mapping. In this way we come to a generalization of the developing process sketched in art. 1.

It should be noticed, however, that this mapping of vectors, which can always be moved parallel to themselves in each local E_n, does not uniquely determine a mapping of *points* of an E_n upon the E_n at a point in the immediate neighborhood. Various assumptions are still allowed. We can for instance map each E_n upon the "next" by mapping the point $P(\xi^\varkappa)$ upon the corresponding point $P_1(\xi^\varkappa + d\xi^\varkappa)$ of the X_n. Another way is that of mapping P_1 on that point of the E_n at P which corresponds to $-d\xi^\varkappa$. This last way, indicated by CARTAN, allows a simple interpretation of the torsion.

3. Torsion. The functions $\Gamma^\varkappa_{\mu\lambda}$ are not necessarily symmetrical in μ and λ. From the transformation formulas it can be shown that $\Gamma^\varkappa_{\mu\lambda} - \Gamma^\varkappa_{\lambda\mu}$ is a tensor. We write

$$S_{\mu\lambda}^{\cdot\cdot\varkappa} = \tfrac{1}{2}(\Gamma^\varkappa_{\mu\lambda} - \Gamma^\varkappa_{\lambda\mu}) = \Gamma^\varkappa_{[\mu\lambda]} = -S_{\lambda\mu}^{\cdot\cdot\varkappa}.$$

The tensor character follows from the formula ($d_1\xi^\varkappa$ and $d_2\xi^\varkappa$ are two different line elements)

$$\delta_2 d_1 \xi^\varkappa - \delta_1 d_2 \xi^\varkappa = 2\, d_1 \xi^\mu d_2 \xi^\lambda S_{\mu\lambda}^{\cdot\cdot\varkappa},$$

which shows that $\delta_2 d_1 \xi^\varkappa$, the covariant differential of $d_1\xi^\mu$ in the $d_2\xi^\varkappa$-direction is not equal to $\delta_1 d_2 \xi^\varkappa$. This also means that ($\varepsilon$ being a small constant)

$$\int \delta\, d\, \xi^\varkappa = 2\varepsilon f^{[\mu\lambda]} S_{\mu\lambda}^{\cdot\cdot\varkappa},$$

taken along an infinitesimal circuit determined by the infinitesimal bivector $\varepsilon f^{[\mu\lambda]}$; therefore, we see that the tensor $S_{\mu\lambda}^{\cdot\cdot\varkappa}$ measures the deviation of the point P from its original position after the local E_n has been mapped consecutively in the sense of CARTAN on the E_n along the circuit until it returns. It returns to its original position if

$$S_{\mu\lambda}^{\cdot\cdot\varkappa} = 0.$$

In this case we call the connection *symmetrical*. It may be denoted by A_n. In the other case $S_{\mu\lambda}^{\cdot\cdot\varkappa} \neq 0$, we say that the L_n at P has torsion, and $S_{\mu\lambda}^{\cdot\cdot\varkappa}$ is called the *torsion tensor* [1]. When $S_{\mu\lambda}^{\cdot\cdot\varkappa} = S_{[\mu} A^\varkappa_{\lambda]}$, the connection is called *semisymmetrical* [2].

---

[1] CARTAN: 1922 (6). — Here also the name torsion. Conception introduced by EDDINGTON: 1921 (1). — See SCHOUTEN: 1924 (5).

[2] SCHOUTEN: 1922 (1) — 1924 (5) p. 73 — 1926 (2). — FRIEDMANN-SCHOUTEN: 1924 (7).

The $\Gamma_{\mu\lambda}^{\varkappa}$ is the case of a general L_n can be expressed in terms of $S_{\mu\lambda}^{\cdot\cdot\varkappa}$ and the covariant derivative of an arbitrary symmetrical tensor $g_{\mu\lambda}=g_{\lambda\mu}$. We find, if the CHRISTOFFEL symbol refers to $g_{\lambda\mu}$:

$$\Gamma_{\mu\lambda}^{\varkappa}=\begin{Bmatrix}\varkappa\\ \mu\lambda\end{Bmatrix}+g_{\nu(\mu}\nabla_{\lambda)}g^{\varkappa\nu}-\tfrac{1}{2}g_{\lambda\pi}g_{\mu\varrho}g^{\varkappa\sigma}\nabla_\sigma g^{\pi\varrho}+S_{\mu\lambda}^{\cdot\cdot\varkappa}-2g^{\varkappa\pi}g_{\varrho(\mu}S_{\lambda)\pi}^{\cdot\cdot\varrho}.$$ [1]

4. WEYL connection. Of special interest are those affine connections in an X_n for which there exists a symmetrical pseudotensor $\mathfrak{g}_{\lambda\mu}=g_{\lambda\mu}\,\mathfrak{e}$ of rank n and class 1, defined for a function τ independent of the parameter $\partial_\varkappa\xi^{\varkappa'}$ and for which the covariant differential vanishes:

$$\nabla_\mu\mathfrak{g}_{\lambda\nu}=0.$$

This gives for $\Gamma_{\mu\lambda}^{\varkappa}$ the condition

$$\Gamma_{\mu\lambda}^{\varkappa}=\begin{Bmatrix}\varkappa\\ \mu\lambda\end{Bmatrix}'-\tfrac{1}{2}(\Gamma_\mu A_\lambda^\varkappa+\Gamma_\lambda A_\mu^\varkappa-g^{\varkappa\nu}g_{\mu\lambda}\Gamma_\nu),$$

where $\mathfrak{g}^{\lambda\mu}$ is determined by the relation $\mathfrak{g}^{\lambda\nu}\mathfrak{g}_{\mu\nu}=A_\mu^\lambda$ and $\begin{Bmatrix}\varkappa\\ \mu\lambda\end{Bmatrix}'$ is the CHRISTOFFEL symbol constructed with the pseudotensor $\mathfrak{g}_{\mu\lambda}$.

For $g_{\lambda\mu}$ we find

$$\nabla_\mu g_{\lambda\nu}=\nabla_\mu\mathfrak{e}^{-1}\mathfrak{g}_{\lambda\nu}=-\mathfrak{e}^{-2}(\nabla_\mu\mathfrak{e})\,\mathfrak{g}_{\lambda\nu}=\Gamma_\varkappa\overset{\varkappa}{e}_\mu g_{\lambda\nu}.$$

If we transform ξ° into $\xi^{\circ\prime}$, the scalar $\mathfrak{e}$ is changed to $\mathfrak{e}'$, and we get for the new tensor $g'_{\lambda\mu}$

$$g'_{\lambda\mu}=\tau g_{\lambda\mu}.$$

At the same time

$$\Gamma'_\lambda=\Gamma_\lambda+\partial_\lambda\ln\tau.$$

As long as ξ° does not change, the Γ_λ behaves like a vector, which we write $-Q_\lambda$. Then the Q_λ is changed under a transformation of the ξ° as follows

$$'Q_\lambda=Q_\lambda-\partial_\lambda\ln\tau$$

and

$$\nabla_\mu g_{\lambda\nu}=-Q_\mu g_{\lambda\nu},$$

(4.1) $$\Gamma_{\mu\lambda}^{\varkappa}=\begin{Bmatrix}\varkappa\\ \mu\lambda\end{Bmatrix}+Q_{(\mu}A_{\lambda)}^{\varkappa}-\tfrac{1}{2}g^{\nu\varkappa}Q_\nu g_{\mu\lambda};\quad \begin{Bmatrix}\varkappa\\ \mu\lambda\end{Bmatrix}\text{ belongs to } g_{\lambda\nu}.$$

This displacement determines a WEYL connection. It is determined by a pseudotensor $\mathfrak{g}_{\lambda\mu}=\mathfrak{g}_{\mu\lambda}$ of rank n and class $+1$, satisfying $\nabla_\mu\mathfrak{g}_{\lambda\nu}=0$ and by giving the Γ_λ in an arbitrary manner.

Another way of defining this connection without introducing the notion of a pseudotensor is by postulating immediately that a tensor $g_{\lambda\mu}=g_{\mu\lambda}$ exists for which the covariant derivative $\nabla_\mu g_{\lambda\nu}$ breaks up into a product $-Q_\mu g_{\lambda\nu}$. It can then be shown that the tensor $g_{\lambda\mu}$ is determined but for a factor τ: $'g_{\lambda\mu}=\tau g_{\lambda\mu}$ and that the $\Gamma_{\mu\lambda}^{\varkappa}$ take the same form as in (4.1)[2].

[1] For general L_n see also GOŁAB: 1930 (12).

[2] WEYL: 1918 (2), (3). For the method with pseudotensors see SCHOUTEN-HLAVATÝ: 1929 (2). Further literature SCHOUTEN: 1929 (5). — EISENHART: 1927 (1). — HLAVATÝ: 1928 (9) — 1929 (12). — GUGINO: 1933 (5). — CARTAN: 1926 (7).

5. Metrical connection. When a tensor $g_{\lambda\mu} = g_{\mu\lambda}$ of rank n exists[1], for which
$$\nabla_\mu g_{\lambda\nu} = 0,$$
we call the connection *metrical* because we can introduce this tensor $g_{\lambda\nu}$ as fundamental tensor of a metric. The local E_n at each point becomes a euclidean space R_n and parallel displacement is equivalent to the mapping of an R_n orthogonally on an R_n in a neighboring point. There is still a considerable degree of freedom in this mapping.

Two cases are of importance: the case without torsion and the case with torsion. When the torsion is zero
$$S_{\mu\lambda}^{\cdot\cdot\varkappa} = 0,$$
we have a RIEMANNian manifold V_n, because we can show that
$$\Gamma_{\mu\lambda}^{\varkappa} = \Gamma_{\lambda\mu}^{\varkappa} = \left\{ {\varkappa \atop \mu\lambda} \right\},$$
where $\left\{ {\varkappa \atop \mu\lambda} \right\}$ are the CHRISTOFFEL symbols of the second kind belonging to $g_{\lambda\mu}$; the displacement of a vector becomes that of LEVI-CIVITA. The equation $\nabla_\mu g_{\lambda\nu} = 0$ then becomes the identity of RICCI for a RIEMANNian manifold[2].

The case with torsion can also be obtained by introducing into the X_n n independent contravariant vector fields $\underset{i}{v}^\nu$, $i = 1, 2, \ldots, n$ and defining with the aid of these vectors a fundamental tensor $g^{\lambda\mu} = \pm \sum \underset{i}{v}^\lambda \underset{i}{v}^\mu$ (summed on i) with respect to which they are mutually orthogonal unit vectors[3].

6. Curvature. In euclidean space a vector always returns to its original position after parallel transportation along a closed curve. This is not necessarily the case in an L_n. We may therefore use the difference between a vector before and after parallel displacement along a closed curve as a measure of the *curvature* of an L_n at a point P. The formula for an infinitesimal circuit along an E_2-element at P, measured by $f^{\varkappa\lambda} d\sigma$, $f^{\varkappa\lambda}$ being a simple bivector and $d\sigma$ an affine measure for the area, is[4]
$$D v^\varkappa = f^{\nu\mu} R_{\nu\mu\lambda}^{\cdot\cdot\cdot\varkappa} v^\lambda d\sigma,$$
$$D w_\lambda = -f^{\nu\mu} R_{\nu\mu\lambda}^{\cdot\cdot\cdot\varkappa} w_\varkappa d\sigma,$$
where $R_{\nu\mu\lambda}^{\cdot\cdot\cdot\varkappa}$ is the *curvature tensor* (or RIEMANN-CHRISTOFFEL tensor)
$$R_{\nu\mu\lambda}^{\cdot\cdot\cdot\varkappa} = -2\partial_{[\nu}\Gamma_{\mu]\lambda}^{\varkappa} - 2\Gamma_{[\nu|\pi|}^{\varkappa}\Gamma_{\mu]\lambda}^{\pi},$$

[1] For the case of rank $<n$ see BORTOLOTTI: 1931 (7).

[2] We do not discuss RIEMANNian manifolds in any detail. See e. g. BERWALD: 1927 (9). — CARTAN: 1925 (8) — 1928 (16). — DUSCHEK-MAYER: 1930 (20). For the conditions that the $\Gamma_{\mu\lambda}^{\varkappa}$ may be written as CHRISTOFFEL symbols see EISENHART: 1927 (1) § 29. — GRAUSTEIN: 1930 (7).

[3] For other types of L_n see KUNII: 1931 (37). — NOVOBATZKY: 1931 (27). — STRANEO: 1932 (27). — NALLI: 1931 (24). — FERNANDES: 1931 (21).

[4] An exact derivation e. g. in SCHLESINGER: 1928 (8).

where $|\pi|$ means that the π is not to be included in the alternation. For this tensor we have the *first identity*

$$\text{(I)} \qquad R_{\nu\mu\lambda}^{\cdot\cdot\cdot\varkappa} = -R_{\mu\nu\lambda}^{\cdot\cdot\cdot\varkappa}, \quad \text{or} \quad R_{(\nu\mu)\lambda}^{\cdot\cdot\cdot\varkappa} = 0,$$

which follows from the definition.

The corresponding formulas for higher order quantities are of the following form

$$D v_{\mu}^{\cdot\varkappa\lambda} = \oint^{\varrho\sigma} d\sigma (R_{\varrho\sigma\tau}^{\cdot\cdot\cdot\varkappa} v_{\mu}^{\cdot\tau\lambda} + R_{\varrho\sigma\tau}^{\cdot\cdot\cdot\lambda} v_{\mu}^{\cdot\varkappa\tau} - R_{\varrho\sigma\mu}^{\cdot\cdot\cdot\tau} v_{\tau}^{\cdot\varkappa\lambda}).$$

Related are the formulas for the application of $\nabla_{[\nu}\nabla_{\mu]}$, the alternating part of the second covariant derivative, e. g.

$$2\nabla_{[\nu}\nabla_{\mu]} w_{\lambda} = R_{\nu\mu\lambda}^{\cdot\cdot\cdot\varkappa} w_{\varkappa} - 2S_{\nu\mu}^{\cdot\cdot\varkappa} \nabla_{\varkappa} w_{\lambda},$$

in which however a term in $S_{\mu\lambda}^{\cdot\cdot\varkappa}$ appears.

For a pseudoscalar we find

$$D\mathfrak{p} = +2\mathfrak{p} \oint^{\nu\mu} \partial_{[\mu}\Gamma_{\nu]} d\sigma.$$

7. Integrability. When $D v^{\varkappa} = 0$ for every circuit and every vector we must have

$$R_{\nu\mu\lambda}^{\cdot\cdot\cdot\varkappa} = 0.$$

In this case parallel transportation of a vector and of every tensor from one point of L_n to another is independent of the curve along which the displacement takes place. It is possible to define at every point a vector (tensor) parallel to a given vector (tensor). There exists *teleparallelism* or *absolute parallelism*, as in euclidean space. Such displacements are called *integrable*.

For pseudotensors, integrability exists if

$$\partial_{[\mu}\Gamma_{\nu]} = 0.$$

In the special case of densities, this means

$$\partial_{[\mu}\Gamma_{\lambda]\varkappa}^{\varkappa} = R_{\mu\lambda\varkappa}^{\cdot\cdot\cdot\varkappa} = 0,$$

so that integrability for tensors also implies integrability for densities. But it implies more. A volume is a scalar density. Hence we see that the equation $R_{\nu\mu\lambda}^{\cdot\cdot\cdot\lambda} = 0$ expresses the fact that teleparallelism exists for volumes. Such an L_n is called *equivoluminar*[1]. For such a manifold it must be possible to select a scalar density $\mathfrak{p}$ in such a way that $\delta\mathfrak{p} = 0$ for all directions $d\xi^{\varkappa}$.

A RIEMANNian manifold V_n with $R_{\nu\mu\lambda}^{\cdot\cdot\cdot\varkappa} = 0$ has the property of admitting n mutually orthogonal gradient fields $\overset{i}{u}_{\mu} = \nabla_{\mu}\overset{i}{\varphi}, i = 1, 2, \ldots, n$. These $\overset{i}{\varphi}$ can be taken as a new CARTESian coordinate system. We call the connection *euclidean*, and we denote it by R_n. It is applicable to euclidean space.

[1] VEBLEN: 1922 (5). — SCHOUTEN: 1924 (5) p. 89.

A symmetrical manifold A_n with $R_{\nu\mu\lambda}^{\cdot\cdot\cdot\,\varkappa} = 0$ is called *affine euclidean.* It is applicable to the space E_n belonging to the affine group in the sense of KLEIN[1].

A metrical manifold with $R_{\nu\mu\lambda}^{\cdot\cdot\cdot\,\varkappa} = 0$ without torsion is euclidean. With torsion it has been the subject of many investigations by WEITZENBÖCK, VITALI and others. EINSTEIN proposed it for $n = 4$ in 1928 as a space-time of relativity[2].

It is known as a RIEMANNian *manifold with torsion,* or RIEMANNian manifold admitting absolute parallelism.

It can be obtained by introducing a fundamental symmetrical tensor $g_{\lambda\mu}$ by means of n contravariant vector fields (see Ch. II, art. 5), and by defining a parallel displacement which carries every vector at a point P over into a vector at another point P' with exactly the same length and position with respect to the unit vectors of the n congruences. A simple example can be constructed by drawing meridians and parallels on a sphere and by defining a parallel displacement which brings every vector making an angle α with the meridian into a similarly situated vector of equal length at another point[3]. It is clear that this is not a displacement of LEVI-CIVITA. It also indicates how this connection can be mapped on a RIEMANNian manifold with a given system of n mutually orthogonal congruences[4]. CLIFFORD parallelism in elliptic space can also be interpreted as a connection of this type. It deserves mention that the teleparallelism in this connection is independent of the metric, as it can be defined with n contravariant vector fields. This teleparallelism is unchanged if the n vector fields are replaced by n linear combinations with constant coefficients. For its application to group theory see this Chapter, art. 10.

8. Some identities[5]**.** Apart from the first identity (art. 5) we have, for the curvature tensor, the following identities in L_n:

(II) $$R_{[\nu\mu\lambda]}^{\cdot\cdot\cdot\,\varkappa} = -2\nabla_{[\nu}S_{\mu\lambda]}^{\cdot\cdot\,\varkappa} + 4S_{[\nu\mu}^{\cdot\cdot\,\pi}S_{\lambda]\pi}^{\cdot\cdot\,\varkappa}.$$

If $\nabla_\mu g^{\lambda\varkappa} = Q_\mu^{\cdot\lambda\varkappa}$ and $R_{\nu\mu\lambda}^{\cdot\cdot\cdot\,\pi} g_{\pi\varkappa} = R_{\nu\mu\lambda\varkappa}$, a third identity exists:

(III) $$R_{\nu\mu(\lambda\varkappa)} = -\nabla_{[\nu}Q_{\mu]\lambda\varkappa} - S_{\nu\mu}^{\cdot\cdot\,\pi}Q_{\pi\lambda\varkappa},$$

so that for every symmetrical connection $R_{[\nu\mu\lambda]}^{\cdot\cdot\cdot\,\varkappa} = 0$ and for every RIEMANNian connection, $R_{\nu\mu(\lambda\varkappa)} = 0$.

---

[1] SCHOUTEN: 1924 (5) Ch. IV.

[2] WEITZENBÖCK: 1921 (4) No. 18 — 1923 (10) p. 317. — EINSTEIN: 1928 (2) — 1930 (11). — See the comprehensive articles of BORTOLOTTI: 1929 (8). — CARTAN: 1930 (9). — EISENHART: 1933 (7). — See further REICHENBACH: 1929 (21). — BORTOLOTTI: 1931 (4). — THOMAS: 1930 (1). — ZAYCOFF: 1931 (21). — LANCZOS: 1931 (33). — TAMM: 1929 (16). — ROBERTSON: 1932 (21). — VITALI: 1932 (20). — SEN: 1931 (25) — and our Ch. II art. 10. Comp. also HOSOKAWA: 1932 (11).

[3] CARTAN: 1923 (1) p. 404 — 1924 (3) p. 301. — Comp. ANDERSON: 1929 (20).

[4] LEVI-CIVITA: 1929 (10). [5] SCHOUTEN: 1924 (5), Ch. II.

If both identities exist, that is, in a RIEMANNian manifold, we can find by purely algebraical computation a fourth identity

$$\text{(IV)} \qquad R_{\nu\mu\lambda\varkappa} = R_{\lambda\varkappa\nu\mu}.$$

For an L_n the curvature tensor has $\frac{1}{2}n^3(n-1)$ linearly independent components. For a V_n the number reduces to $\frac{1}{12}n^2(n^2-1)$.

The identity of BIANCHI for an L_n is

$$\nabla_{[\mu} R_{\nu\pi]\lambda}^{\cdot\cdot\cdot\,\varkappa} = \overset{+}{-}2 S_{[\nu\pi}^{\cdot\cdot\,\sigma} R_{\mu]\sigma\lambda}^{\cdot\cdot\cdot\,\varkappa}.$$

By contraction we find from this identity that for a V_n

$$\nabla_\mu G^{\mu}_{\cdot\lambda} = 0, \quad \text{if} \quad G_{\mu\lambda} = R_{\mu\lambda} - \tfrac{1}{2} R g_{\mu\lambda},$$

where

$$R_{\mu\lambda} = R_{\nu\mu\lambda}^{\cdot\cdot\cdot\,\nu}, \qquad R = R_{\mu\lambda} g^{\mu\lambda}.$$

In a V_n the $R_{\mu\lambda}$ is symmetrical. In a general L_n there is a symmetrical and an alternating part to it, which fact has occasionally been used for relativity, where a symmetrical tensor and a bivector have to define the gravitational and the electromagnetic field.

9. Non-holonomic systems. So far we have considered only geometrical properties referred to holonomic systems. If we now introduce non-holonomic measuring vectors[1], we can express the displacement of a contravariant vector in the L_n in this way

$$\begin{aligned} \nabla_j v^k &= A_{j\nu}^{\mu k} \nabla_\mu v^\nu = A_{j\nu}^{\mu k} \partial_\mu v^\nu + A_{j\nu}^{\mu k} \Gamma_{\mu\lambda}^{\nu} v^\lambda = \\ &= \partial_j v^k + A_{ij\nu}^{\lambda\mu k} \Gamma_{\mu\lambda}^{\nu} v^i + v^i A_\lambda^k \partial_j A_i^\lambda, \\ &= \partial_j v^k + \Gamma_{ji}^k v^i, \end{aligned}$$

where

$$\partial_j = A_j^\mu \partial_\mu$$

$$\Gamma_{ji}^k = A_{ij\nu}^{\lambda\mu k} \Gamma_{\mu\lambda}^{\nu} + A_\lambda^k \partial_j A_i^\lambda = A_{ij\nu}^{\lambda\mu k} \Gamma_{\mu\lambda}^{\nu} - A_i^\lambda \partial_j A_\lambda^k.$$

We can write in a similar way

$$\nabla_j w_i = \partial_j w_i - \Gamma_{ij}^k w_k.$$

The Γ_{ij}^k can be taken as the parameters of displacement in the non-holonomic system. We have

$$\Gamma_{ij}^k \overset{*}{=} \nabla_j \underset{i}{e}^k; \qquad \Gamma_{ij}^k \overset{*}{=} \nabla_j \overset{k}{e}_i.$$

When the $\Gamma_{\mu\lambda}^{\varkappa}$ are symmetrical, the Γ_{ij}^k need not be symmetrical. As

$$\Gamma_{[ij]}^k = S_{ij}^{\cdot\cdot\,k} - A_{ij}^{\mu\lambda} \partial_{[\mu} A_{\lambda]}^k,$$

we see that the measuring vectors are holonomic when and only when $\Gamma_{[ji]}^k = S_{ji}^{\cdot\cdot\,k}$. In the non-holonomic case the $\Gamma_{[ji]}^k$ have no tensor character.

---

1 SCHOUTEN: 1929 (4).

If we write
$$\Omega^{k}_{ij} = -A^{\mu\lambda}_{ij}\partial_{[\mu}A^{k}_{\lambda]},$$
we have $S_{ji}^{\cdot\cdot k} = \Gamma^{k}_{[ji]} + \Omega^{k}_{ij}$. The Ω^{k}_{ij} are called the *anholonomic parameters*. They form a geometrical object, not a quantity.

The non-holonomic components of the curvature tensor take the following form
$$R_{ijk}^{\cdot\cdot\cdot l} = -2\partial_{[i}\Gamma^{l}_{j]k} - 2\Gamma^{l}_{[i|m|}\Gamma^{m}_{j]k} - 2\Omega^{m}_{ij}\Gamma^{l}_{mk}.$$

An application to RIEMANNian geometry can be made by introducing as non-holonomic measuring vectors the unit tangent vectors $\underset{i}{i}^{\nu}$, $\overset{k}{i}_{\lambda}$ along an orthogonal ennuple (RICCI's λ^{i}_{k}, $\lambda_{h|j}$). Then we find
$$\Gamma^{k}_{ij} \overset{*}{=} \nabla_{j}\underset{i}{i}^{k}, \qquad \Gamma^{k}_{ik} = 0,$$

which shows that the Γ^{k}_{ij} are the *rotation coefficients*[1] of RICCI, belonging, in RICCI's notation, to the ennuple, or:
$$g_{hk}\Gamma^{k}_{ij} \overset{*}{=} (\text{Ricci notation}) \overset{*}{=} \gamma_{ihj} = -\gamma_{hij}.$$

With these non-holonomic displacements in V_n also deal some papers by CISOTTI and PASTORI[2].

In the RIEMANNian connection with torsion (art. 7), the fields $\underset{i}{v}^{\varkappa}$ also build a non-holonomic system of reference. They may be used for a holonomic system as soon as the corresponding covariant fields $\overset{i}{v}_{\lambda}$ form X_{n-1}.[3]

10. Transformation groups. The transformations of a finite continuous simple group in n parameters $\xi^{\varkappa}$ can be represented as points in an L_n, in which two kinds of RIEMANNian connections with torsion can be defined. If T_{ξ} represents the general transformation of the group, then the parameters of the infinitesimal transformation $T_{\xi}^{-1}T_{\xi+d\xi}$ define the n contravariant vectorfields of the first connection, and those of $T_{\xi+d\xi}\,T_{\xi}^{-1}$ the vectorfields of the second connection. The components of the torsion tensor $S_{\lambda\mu}^{\cdot\cdot\varkappa}$ are equal to the constants of the structure c_{ijk} of LIE.

For such connections the geodesics coincide with those of the RIEMANNian connection with the same definite ds^2. An example is elliptic space of 3 dimensions, in which the connections with torsion are those with CLIFFORD parallelism[4].

---

[1] E. g. RICCI-LEVI CIVITA: Math. Ann. Vol. 54 (1901).

[2] Comp. PASTORI: 1930 (17). — INFELD: 1932 (24). — VRANCEANU: 1932 (34).

[3] SCHOUTEN: 1929 (3).

[4] See SCHOUTEN: 1929 (5). — CARTAN: 1927 (13) — 1930 (9); in the latter the literature is given. Also: EISENHART: Proc. Acad. Sci. U. S. A. Vol. 11 (1925) p. 246. — SLEBODZINSKI: 1932 (25), (26). — WHITEHEAD: 1932 (18).

Chapter III.
Connections associated with differential equations.

1. Paths. In a RIEMANNian geometry a geodesic can be defined as a curve generated by a linear element moved parallel to itself in its own direction. This definition can immediately be extended to an L_n. If the linear element is denoted by the contravariant vector $v^\varkappa$ we must express that $v^\mu \nabla_\mu v^\varkappa$ has the direction of $v^\varkappa$. If the geodesic has the equations $\xi^\varkappa = \xi^\varkappa(t)$, we find for its differential equation

(1.1) $$\frac{d\xi^\mu}{dt}\nabla_\mu \frac{d\xi^\varkappa}{dt} = \alpha \frac{d\xi^\varkappa}{dt}, \qquad \alpha \text{ coefficient depending on } \xi^\varkappa$$

or

$$\frac{d^2\xi^\varkappa}{dt^2} + \Gamma^\varkappa_{\mu\lambda}\frac{d\xi^\mu}{dt}\frac{d\xi^\lambda}{dt} = \alpha\frac{d\xi^\varkappa}{dt}.$$

It is possible to find an invariant parameter $s = s(t)$ on the curve

$$s = c_1 + c_2 e^{\int \alpha\, dt}\, dt, \qquad c_1, c_2 \text{ constants},$$

by which the equation of the geodesic takes the form

(1.2) $$\frac{d^2\xi^\varkappa}{ds^2} + \Gamma^\varkappa_{\mu\lambda}\frac{d\xi^\mu}{ds}\frac{d\xi^\lambda}{ds} = 0.$$

This is a system of n differential equations which, in a certain domain of L_n, allow a solution such that through each point passes an integral curve in every one of the ∞^{n-1} directions, and one integral curve passing through two points. It defines, therefore, a system of ∞^{2n-2} geodesics, also called *paths*[1].

The most general system of paths is given by the differential equation

(1.3) $$\frac{d^2\xi^\varkappa}{dt^2} = f^\varkappa\left(\xi^\lambda, \frac{d\xi^\mu}{dt}\right),$$

where the $f^\varkappa$ are homogeneous of the second degree in $d\xi^\mu/dt$.[2] In this case, however, the Γ depend, as a rule, on $d\xi^\varkappa/dt$, a case which we do not discuss in detail.

It is now possible to begin the investigation with a system (1.1) of differential equations, and to define the connection by its coefficients $\Gamma^\varkappa_{\mu\lambda}$. Instead of letting the connection define the paths, the paths can be made to define the connection. In this case, however, the paths

---

[1] EISENHART-VEBLEN: 1922 (3). — See EISENHART: 1927 (1). — Also WHITEHEAD: 1932 (19).

[2] DOUGLAS: 1928 (1). — ROWE: 1932 (29). — RASCHEWSKY: 1932 (10). — Generalization of the system of equations (1.3) in DOUGLAS: 1931 (15).

always define a symmetrical connection $\Gamma^{\varkappa}_{\mu\lambda} = \Gamma^{\varkappa}_{\lambda\mu}$ as the torsion does not affect the paths. To one system of paths belongs therefore an infinity of L_n.

2. Projective transformations. A system (1.1) of paths does not even define uniquely one connection A_n. Indeed, the transformation

$$(2.1) \qquad {}'\Gamma^{\varkappa}_{\mu\lambda} = \Gamma^{\varkappa}_{\mu\lambda} + A^{\varkappa}_{\lambda} p_{\mu} + A^{\varkappa}_{\mu} p_{\lambda} = \Gamma^{\varkappa}_{\mu\lambda} + 2 A^{\varkappa}_{(\lambda} p_{\mu)},$$

where p_μ is an arbitrary covariant vector, leaves the equations (1.2) invariant though it may change the parameter s. The transformation fails to change the parameter on the paths only if $p_\mu d\xi^\mu = 0$, that is, if the E_{n-1} of p_μ contains the path direction.

For an asymmetric connection a more general transformation preserves paths:

$$(2.2) \qquad {}'\Gamma^{\varkappa}_{\mu\lambda} = \Gamma^{\varkappa}_{\mu\lambda} + p_{\mu} A^{\varkappa}_{\lambda} + q_{\lambda} A^{\varkappa}_{\mu}. \qquad (p_\lambda, q_\lambda \text{ arbitrary vectors})$$

We say that all manifolds A_n with the same paths are *projectively* related, and the transformation (2.1) is called a *projective transformation* of the A_n.[1] The projective geometry of A_n is the theory of geometrical objects defined with respect to these transformations.

The curvature tensor transforms under (2.1) as follows

$$(2.3) \qquad \begin{cases} {}'R_{\nu\mu\lambda}^{\cdot\cdot\cdot\varkappa} = R_{\nu\mu\lambda}^{\cdot\cdot\cdot\varkappa} - 2 p_{[\nu\mu]} A^{\varkappa}_{\lambda} + 2 A^{\varkappa}_{[\nu} p_{\mu]\lambda}, \\ p_{\mu\lambda} = \nabla_{\mu} p_{\lambda} - p_{\mu} p_{\lambda}. \end{cases}$$

This tensor is therefore not invariant under projective transformations. From it, however, we can derive the tensor

$$P_{\nu\mu\lambda}^{\cdot\cdot\cdot\varkappa} = R_{\nu\mu\lambda}^{\cdot\cdot\cdot\varkappa} - 2 P_{[\nu\mu]} A^{\varkappa}_{\lambda} + 2 A^{\varkappa}_{[\nu} P_{\mu]\lambda},$$

$$P_{\mu\lambda} = -\frac{1}{n^2-1}(n R_{\mu\lambda} + R_{\lambda\mu}), \qquad R_{\mu\lambda} = R_{\nu\mu\lambda}^{\cdot\cdot\cdot\nu},$$

and a verification shows that this tensor is unchanged by a projective transformation. It is called the *projective curvature tensor*, and vanishes identically for $n = 1$, $n = 2$. For $n > 2$ it satisfies the identities:

$$P_{\nu\mu\lambda}^{\cdot\cdot\cdot\varkappa} = -P_{\mu\nu\lambda}^{\cdot\cdot\cdot\varkappa}, \qquad P_{[\nu\mu\lambda]}^{\cdot\cdot\cdot\varkappa} = 0, \qquad \nabla_{[\mu} P_{\nu\pi]\lambda}^{\cdot\cdot\cdot\varkappa} = \frac{1}{n-2} A^{\varkappa}_{[\nu} \nabla_{|\varkappa|} P_{\pi\mu]\lambda}^{\cdot\cdot\cdot\nu},$$

which can be verified from the corresponding identities for the curvature tensor $R_{\nu\mu\lambda}^{\cdot\cdot\cdot\varkappa}$.

The vanishing of the projective curvature tensor for $n > 2$ is the necessary and sufficient condition that the A_n can be changed, by a projective transformation, into a euclidean manifold R_n. Such a manifold is called *projective-euclidean* and its paths pass into the straight lines of the R_n. For $n = 2$, when the projective curvature tensor does not exist, another condition is necessary, namely, that $P_{\mu\lambda}$ (existing for $n = 2$) satisfies

[1] Weyl: 1921 (2). — For condition (2.2) see Hlavatý: 1926 (3) — comp. 1927 (16). — Related is Schouten: 1927 (10).

the condition $\nabla_{[\mu} P_{\lambda]\nu} = 0$. Indeed, a surface in ordinary space cannot, as a rule, be mapped on a plane with the preservation of the geodesics, it has to be of constant curvature. The general theorem can be found by writing down the conditions of integrability of the equations

$$0 = R_{\nu\mu\lambda}^{\cdot\cdot\cdot\varkappa} - 2p_{[\nu\mu]} A_\lambda^\varkappa - 2A_{[\nu}^\varkappa p_{\mu]\lambda},$$

which follow from (2.3) by the assumption that $'R_{\nu\mu\lambda}^{\cdot\cdot\cdot\varkappa} = 0$.[1]

Point transformations which preserve the paths are called *collineations*. The properties of finite continnous groups of collineations have been investigated[2].

3. THOMAS parameters. A geometrical object unaltered by a projective transformation of A_n is

$$\Pi_{\mu\lambda}^\varkappa = \Gamma_{\mu\lambda}^\varkappa - \frac{2}{n+1} A_{(\mu}^\varkappa \Gamma_{\lambda)\nu}^\nu .$$[3]

These $\Pi_{\mu\lambda}^\varkappa$, which satisfy the identity $\Pi_{\mu\varkappa}^\varkappa = 0$, may be considered as the parameters of a displacement, which is uniquely determined by the paths as soon as the coordinate system is fixed. They determine a kind of projective displacement, of which the paths are the solution of the differential equations

$$\frac{d^2\xi^\varkappa}{dp^2} + \Pi_{\mu\lambda}^\varkappa \frac{d\xi^\mu}{dp}\frac{d\xi^\lambda}{dp} = 0,$$

the p being a normalised projective parameter defined but for two constants c_1 and c_2:

$$p = c_1 \int \Delta^{\frac{2}{n+1}} dt + c_2, \qquad \Delta = \operatorname{Det} |\partial_\varkappa \xi^{\varkappa'}|.$$

The Δ enters here, as it does in the definition of densities; it also enters into the transformation equations of the $\Pi_{\mu\lambda}^\varkappa$ when we pass from $(\varkappa)$ to $(\varkappa')$, which can be written

$$\Pi_{\mu'\lambda'}^{\varkappa'} = A_{\varkappa\mu'\lambda'}^{\varkappa'\mu\lambda} \Pi_{\mu\lambda}^\varkappa + A_\nu^{\varkappa'} \partial_{\mu'} A_{\lambda'}^\nu - \frac{2}{n+1} A_{(\lambda'}^{\varkappa'} \partial_{\mu')} \ln \Delta .$$

When $\Delta = 1$ the Π transform like the Γ and p is independent of the coordinate system. This is the *equiprojective* case[4].

This occurence of the Δ shows that the projective parameter p depends not only on the curve, but also on the choice of original variables. The problem of finding projective equivalence of A_n can be reduced to the study of the integrability conditions of the transformation equations of the $\Pi_{\mu\lambda}^\varkappa$.[5] For a further treatment of this subject we refer to EISENHART's book[6].

---

[1] See also SCHOUTEN: 1924 (5), Ch. IV; SCHOUTEN: 1926 (2); 1927 (12).
[2] KNEBELMAN: 1928 (17); EISENHART: 1927 (1) p. 127.
[3] THOMAS: 1925 (6); 1926 (3).
[4] THOMAS: 1925 (2).
[5] VEBLEN-THOMAS: 1926 (8).
[6] EISENHART: 1927 (1) Ch. III.

4. Conformal transformations. Closely related in its formal apparatus is the theory of *conformal* transformations of a RIEMANNian manifold V_n

$$'g_{\mu\lambda} = \sigma g_{\mu\lambda}, \quad \sigma = \sigma(\xi^\varkappa).$$

In this case we have for the CHRISTOFFEL symbols of the second kind:

$$(4.1) \qquad \begin{aligned} '\Gamma^\varkappa_{\mu\lambda} &= \Gamma^\varkappa_{\mu\lambda} + A^\varkappa_{(\mu} s_{\lambda)} - \tfrac{1}{2} g^{\varkappa\nu} g_{\mu\lambda} s_\nu, \\ s_\lambda &= \partial_\lambda \ln\sigma. \end{aligned}$$

Such conformal transformations leave the angle between two vectors unaltered. The conformal theory of V_n is the theory of the geometrical objects defined with respect to these transformations.

The curvature tensor transforms under (4.1) as follows:

$$'R_{\nu\mu\lambda}^{\cdot\cdot\cdot\varkappa} = R_{\nu\mu\lambda}^{\cdot\cdot\cdot\varkappa} - 2 g_{\lambda[\nu} s_{\mu]\pi} g^{\pi\varkappa},$$

$$s_{\mu\lambda} = 2 \nabla_\mu s_\lambda - s_\mu s_\lambda + \tfrac{1}{2} s_\varkappa s^\varkappa g_{\mu\lambda}, \qquad s^\varkappa = g^{\varkappa\lambda} s_\lambda.$$

From it can be derived the following tensor invariant under conformal transformations:

$$C_{\nu\mu\lambda}^{\cdot\cdot\cdot\varkappa} = R_{\nu\mu\lambda}^{\cdot\cdot\cdot\varkappa} - \frac{4}{n-2} g_{\lambda(\nu} L_{\mu)}^{\cdot\,\varkappa},$$

$$L_{\mu\lambda} = -R_{\mu\lambda} + \frac{1}{2(n-1)} R g_{\mu\lambda}, \qquad R = g^{\mu\lambda} R_{\varkappa\mu\lambda}^{\cdot\cdot\cdot\varkappa}.$$

This is the conformal curvature tensor and it vanishes identically for $n = 1, 2, 3$; for $n > 3$, it satisfies the identities

$$C_{\nu\mu\lambda}^{\cdot\cdot\cdot\varkappa} = -C_{\mu\nu\lambda}^{\cdot\cdot\cdot\varkappa}, \qquad C_{[\nu\mu\lambda]}^{\cdot\cdot\cdot\varkappa} = 0, \qquad C_{\nu\mu\lambda\varkappa} = -C_{\nu\mu\varkappa\lambda};$$

here also

$$C_{\nu\mu\lambda\varkappa} = C_{\lambda\varkappa\nu\mu}.$$

The vanishing of the conformal curvature tensor for $n > 3$ is the necessary and sufficient condition that the V_n can be mapped on an R_n by a conformal transformation[1]. Such a manifold is called *conformal-euclidean.* For $n = 3$, when the conformal curvature tensor vanishes, the condition is that $L_{\mu\lambda}$ (existing for $n = 3$) shall satisfy the condition of COTTON that $\nabla_{[\mu} L_{\lambda]\nu} = 0$. A V_2 can always be conformally mapped on an R_2 in as many ways as there are analytical functions of a complex variable.

A geometrical object unaltered by a conformal transformation is

$$Z^\varkappa_{\mu\lambda} = \Gamma^\varkappa_{\mu\lambda} - \frac{2}{n} A^\varkappa_{(\mu} \Gamma^\nu_{\lambda)\nu} + \frac{1}{n} g^{\varkappa\nu} g_{\mu\lambda} \Gamma^\pi_{\nu\pi}$$

for which $Z^\varkappa_{\mu\varkappa} = 0$. The $Z^\varkappa_{\mu\lambda}$ may be taken as parameters of a displacement, which is uniquely determined as soon as the metric and the coordinate system is given.

[1] For literature see SCHOUTEN: 1924 (5) p. 170. — See also 1927 (12).

It is possible to build up a theory of conformal invariants in V_n starting with the remark that the quantity

$$G_{\lambda\mu} = g_{\lambda\mu} |g_{\lambda\mu}|^{-1/n}, \qquad |g_{\lambda\mu}| = \text{Determinant of } g_{\lambda\mu}$$

behaves like a tensor density of weight $-2/n$ which is independent of σ. The theory of conformal invariants thus becomes a theory of invariants of tensor densities[1].

It is not possible to get a non-trivial projective transformation for a V_n which is at the same time conformal. Then we need

$$A^{\varkappa}_{(\mu} s_{\lambda)} - \tfrac{1}{2} s^{\varkappa} g_{\mu\lambda} = 2 A^{\varkappa}_{(\mu} p_{\lambda)}$$

or

$$s_\lambda = \frac{2(n+1)}{n} p_\lambda = -\frac{4}{n-2} p_\lambda,$$

which does not give acceptable values for n. Indeed, a V_n is fully determined by its geodesic lines and specification of its fundamental tensor but for a factor[2].

5. Normal coordinates. The equations (1.2) of the paths enable us to define a special set of coordinates at each point of X_n. The integral curve through a point $P(\overset{}{\underset{0}{\xi}}{}^{\varkappa})$ in direction $\underset{0}{v}{}^{\varkappa} = d\underset{0}{\xi}{}^{\varkappa}/ds$ has an equation of the form

$$\xi^{\varkappa} - \underset{0}{\xi}{}^{\varkappa} = \underset{0}{v}{}^{\varkappa} s + \frac{1}{2}\left(\frac{d^2 \xi^{\varkappa}}{ds^2}\right)_0 s^2 + \frac{1}{6}\left(\frac{d^3 \xi^{\varkappa}}{ds^3}\right)_0 s^3 + \cdots.$$

The coefficients of this series, which we suppose to be convergent, can be found by means of (1.2) and its derived equations:

$$\frac{d^3 \xi^{\varkappa}}{ds^3} + \Gamma^{\varkappa}_{\mu_1\mu_2\mu_3} \frac{d\xi^{\mu_1}}{ds} \frac{d\xi^{\mu_2}}{ds} \frac{d\xi^{\mu_3}}{ds} = 0,$$

$$\cdots\cdots\cdots\cdots\cdots\cdots\cdots$$

$$\frac{d^p \xi^{\varkappa}}{ds^p} + \Gamma^{\varkappa}_{\mu_1\mu_2\ldots\mu_p} \frac{d\xi^{\mu_1}}{ds} \frac{d\xi^{\mu_2}}{ds} \cdots \frac{d\xi^{\mu_p}}{ds} = 0,$$

$$\Gamma^{\varkappa}_{\mu_1\mu_2\ldots\mu_p\lambda} = \partial_{(\lambda} \Gamma^{\varkappa}_{\mu_1\mu_2\ldots\mu_p)} - p\,\Gamma^{\varkappa}_{\nu(\mu_1\mu_2\ldots\mu_{p-1}} \Gamma^{\nu}_{\mu_p)\lambda}.$$

Hence

$$\zeta^{\varkappa} = \xi^{\varkappa} - \underset{0}{\xi}{}^{\varkappa} = \underset{0}{v}{}^{\varkappa} s - \tfrac{1}{2} \overset{0}{\Gamma}{}^{\varkappa}_{\mu\lambda} \underset{0}{v}{}^{\lambda} \underset{0}{v}{}^{\mu} s^2 - \tfrac{1}{6} \overset{0}{\Gamma}{}^{\varkappa}_{\mu\lambda\nu} \underset{0}{v}{}^{\lambda} \underset{0}{v}{}^{\mu} \underset{0}{v}{}^{\nu} s^3 \ldots$$

If we write $\underset{0}{v}{}^{\varkappa} s = \eta^{\varkappa}$, we can, by virtue of the fact that $|\partial\xi^{\varkappa}/\partial\eta^{\lambda}| \neq 0$, invert these equations and get

$$\eta^{\varkappa} = \zeta^{\varkappa} + \tfrac{1}{2} \overset{0}{\Lambda}{}^{\varkappa}_{\lambda\mu} \zeta^{\mu} \zeta^{\lambda} + \tfrac{1}{6} \overset{0}{\Lambda}{}^{\varkappa}_{\mu\lambda\nu} \zeta^{\mu} \zeta^{\lambda} \zeta^{\nu} + \cdots \tag{5.1}$$

where

$$\Lambda^{\varkappa}_{\mu\lambda} = \Gamma^{\varkappa}_{\mu\lambda}$$

$$\Lambda^{\varkappa}_{\mu\lambda\nu} = \Gamma^{\varkappa}_{\mu\lambda\nu} + 3 \Lambda^{\varkappa}_{\pi(\mu} \Gamma^{\pi}_{\lambda\nu)}, \quad \text{etc.}$$

---

[1] THOMAS: 1925 (6) — 1932 (9). — VEBLEN: 1928 (4).

[2] WEYL: 1921 (2).

The $\eta^\varkappa$ form the components of a vector which in the local E_n at P may be taken as the radius vector from P to a point. In RIEMANNian geometry the $\eta^\varkappa$ are sometimes called RIEMANN's normal coordinates. We call them *normal coordinates*[1]. They give a representation of a domain about P in which the series converges on the local E_n at P.

The importance of normal coordinates lies mainly in the property that at P the values $\Gamma^\varkappa_{\mu\lambda}$, expressed in the $\eta^\varkappa$ coordinates, disappear. Indeed, from (5.1):

$$\left[\frac{\partial}{\partial\eta^\mu}\frac{\partial\xi^\varkappa}{\partial\eta^\lambda}\right]_{\eta=0} = \frac{\partial}{\partial\eta^\mu}\left[A^\varkappa_\lambda - \overset{0}{\Gamma}{}^\varkappa_{\mu\lambda}\eta^\mu - \cdots\right] = -\overset{0}{\Gamma}{}^\varkappa_{\mu\lambda},$$

so that according to the transformation formulas for the Γ, when passing from $\xi^\varkappa$ to $\eta^\varkappa$:

$$\overset{0}{\Gamma}{}^{\varkappa'}_{\mu'\lambda'} = 0.$$

This holds only at P and only for the Γ, but not in general for their derivatives. At P however all covariant derivatives of the first order in A_n can be written as ordinary derivatives with respect to $\eta^\varkappa$, e. g. $(\nabla_\mu v_{\lambda\nu})_0 = (\partial v_{\lambda\nu}/\partial\eta^\mu)_0$.

This holds for all coordinates which can be defined at P by means of a series in $\zeta^\varkappa$ which have the first two terms in common with (5.1), in particular, the system

$$'\eta^\varkappa = \zeta^\varkappa + \tfrac{1}{2}\overset{0}{\Gamma}{}^\varkappa_{\mu\lambda}\zeta^\mu\zeta^\lambda.$$

In normal coordinates many proofs are very simple, e. g. those for the second identity of the curvature tensor or for the identity of BIANCHI. Their use in the establishment of existence theorems has been shown by EISENHART, VEBLEN, THOMAS and others. We refer here especially to VEBLEN's book on invariants.

It is possible to construct systems of normal coordinates based on the $\Pi^\varkappa_{\mu\lambda}$ of equi-projective or the $Z^\varkappa_{\mu\lambda}$ of conformal displacements. In such systems the study of the objects of such connection is considerably simplified[2].

Related is a theorem of FERMI, which states that for the case of a V_n there is always a coordinate system in which the CHRISTOFFEL symbols vanish along a curve. It can be shown that this also holds for the $\Gamma^\varkappa_{\mu\lambda}$ of an A_n. This means that corresponding to a curve in the A_n there exists an E_n with the same Γ along the curve[3].

---

[1] VEBLEN: 1922 (4). — VEBLEN-THOMAS: 1923 (6). — For V_n see RIEMANN-WEYL: 1854 (1). — Also HLAVATÝ: 1927 (17). — See also THOMAS: 1929 (7). — RUSE: 1931 (29). — MICHAL: 1931 (12).

[2] THOMAS: 1925 (2) — 1930 (2), (3). — EISENHART: 1927 (1). — See esp. VEBLEN: 1927 (2).

[3] EISENHART: 1927 (1) p. 64. — An extension in WHITEHEAD-WILLIAMS: 1930 (25).

6. Displacements defined by a partial differential equation. The previous displacements were all defined with the aid of systems of ordinary differential equations. An entirely different procedure can be followed if a linear partial differential equation of the second order is defined in the X_n. Let it be:

$$F(\psi) \equiv a^{\nu\mu}\,\partial^2_{\nu\mu}\psi + a^{\nu}\,\partial_{\nu}\psi + a^{\circ}\psi = 0\,, \qquad \partial^2_{\nu\mu} = \partial^2/\partial\,\xi^{\nu}\,\partial\,\xi^{\mu},$$
$$a^{\nu\mu},\ a^{\lambda},\ a^{\circ} \text{ functions of } \xi^{\varkappa}.$$

The left hand side remains invariant under a coordinate transformation of the $\xi^{\varkappa}$. It defines a symmetrical tensor, which we assume to be of rank n

$$g^{\mu\lambda} = a^{\mu\lambda}.$$

It also defines a contravariant and a covariant vector:

$$p^{\varkappa} = a^{\varkappa}_{\cdot} - \frac{1}{\sqrt{g}}\,\partial_{\mu}\left(\sqrt{g}\,g^{\varkappa\mu}\right) \qquad \text{and} \qquad p_{\lambda} = g_{\lambda\varkappa}\,p^{\varkappa}$$

respectively, and a scalar field $\xi^{\circ} = a^{\circ}$, so that $F(\psi) = 0$ takes the form

$$g^{\mu\nu}\,\nabla_{\nu}\,\nabla_{\mu}\psi + p^{\mu}\,\nabla_{\mu}\psi + \xi^{\circ}\psi = 0\,.$$

The equation $F(\psi)=0$ also remains invariant under a transformation

$$'F(\psi) = \tau\,F(\psi), \qquad \tau \text{ arbitrary function of } \xi^{\varkappa},$$

by which the $g_{\mu\lambda}$, $p^{\varkappa}$, ξ° transform according to the formulas

$$'g^{\lambda\mu} = \tau\,g^{\lambda\mu}, \qquad 'p_{\lambda} = p_{\lambda} + \ln\tau\,, \qquad '\xi^{\circ} = \tau\,\xi^{\circ}.$$

The equation $F(\psi) = 0$ therefore determines in an X_n a set of coordinates $\xi^{\varkappa}$, ξ° as in Ch. I, art 8, a pseudotensor $\mathfrak{g}^{\lambda\mu}$ of class 1 and a set of parameters of displacement $\Gamma_{\lambda} = p_{\lambda}$. With the aid of these quantities linear displacements can be defined. If, e. g., we assume X_n to be an A_n, we can take $\nabla_{\mu}\,\mathfrak{g}_{\lambda\nu} = 0$ and define a WEYL connection[1]. We can also use these tensors to construct a projective connection (Ch. V).

This method enables us to build up in X_4 theories of relativity on a wave equation of the SCHRÖDINGER type. In this case we must take for $F(\psi) = 0$ an equation of the hyperbolic type which leads to a pseudotensor of MINKOWSKI signature. Determination of the τ can be obtained by suitable gauging. Like all theories of this kind involving pseudotensors it is contained in the more symmetrical theory which works with homogeneous coordinates (Ch. V).

7. Differential comitants. We cannot here discuss the many papers dealing with the construction of complete systems of differential comitants related to different connections. We refer only to EISENHART'S

[1] STRUIK-WIENER: 1927 (4).

book dealing with tensors in A_n[1] and a paper by THOMAS-MICHAL dealing with tensor densities in V_n and containing a discussion of the literature[2].

For other material on this subject see papers by WEITZENBÖCK and KRAUSS[3].

Chapter IV.
HERMITian connections.

1. HERMITian quantities[4]. The variables $x^{\varkappa}$ of an E_n can be made to run through all complex values. We have then to consider the conjugate complex variable $x^{\bar{\varkappa}}$ of $x^{\varkappa}$, $\bar{\varkappa}, \bar{\lambda}, \ldots = \bar{1}, \bar{2}, \ldots, \bar{n}$. Then it is possible to define quantities with respect to the $x^{\bar{\varkappa}}$ in one to one correspondence with those defined with respect to $x^{\varkappa}$ e. g.,

$$v^{\bar{\varkappa}'} = A^{\bar{\varkappa}'}_{\bar{\varkappa}} v^{\bar{\varkappa}}, \qquad v^{\varkappa'} = A^{\varkappa'}_{\varkappa} v^{\varkappa}; \qquad A^{\bar{\varkappa}'}_{\bar{\varkappa}} \text{ is the conj. to } A^{\varkappa'}_{\varkappa}.$$

A next step is the introduction of quantities in which some indices refer to the $x^{\varkappa}$ and some to the $x^{\bar{\varkappa}}$, as

$$g_{\lambda'\bar{\mu}'} = A^{\lambda\bar{\mu}}_{\lambda'\bar{\mu}'} g_{\lambda\bar{\mu}}, \qquad A^{\lambda\bar{\mu}}_{\lambda'\bar{\mu}'} = A^{\lambda}_{\lambda'} A^{\bar{\mu}}_{\bar{\mu}'}.$$

Such quantities are called HERMITian. To each HERMITian quantity belongs one and only one quantity with complex conjugate components; we denote them through the indices, as in $P_{\mu\bar{\lambda}}^{\cdot\,\cdot\,\bar{\varkappa}} \to P_{\bar{\mu}\lambda}^{\cdot\,\cdot\,\varkappa}$. We can construct HERMITian tensors and densities in a way similar to those defined with real variables. With the definition of symmetry and alternation we must be careful, because a tensor like $v_{\lambda\bar{\mu}\bar{\nu}}$ defines $v_{\bar{\lambda}\mu\nu}$, but not $v_{\bar{\mu}\lambda\bar{\nu}}$. Without an additional assumption we are able, however, to define symmetry for tensors of the second order. Indeed, the equation

$$h_{\lambda\bar{\mu}} = h_{\bar{\mu}\lambda} \quad (\text{e. g. for } n = 2: \quad h_{1\bar{1}} = h_{\bar{1}1}, \quad h_{1\bar{2}} = h_{\bar{2}1}, \quad h_{2\bar{1}} = h_{\bar{1}2}, \quad h_{2\bar{2}} = h_{\bar{2}2}),$$

is fully determined and is preserved under coordinate transformations; the same holds for an alternating tensor of second order:

$$h_{\lambda\bar{\varkappa}} = -h_{\bar{\varkappa}\lambda}.$$

---

[1] EISENHART: 1927 (1). — Further THOMAS: 1929 (7) — 1930 (2), (3).

[2] THOMAS-MICHAL: 1927 (5).

[3] KRAUSS: 1927 (6), and e. g. WEITZENBÖCK: 1932 (32).

[4] SCHOUTEN-VAN DANTZIG: 1929 (6) — 1930 (6). For the purpose of simplicity we write for the conjugate complex of $x^{\varkappa}$ the symbol $x^{\bar{\varkappa}}$, for the conjugate complex of $g_{\lambda\mu}$ the symbol $g_{\bar{\lambda}\bar{\mu}}$, etc. Professor J. A. SCHOUTEN remarks in a letter that this may lead to ambiguities and suggests for the conjugate complex quantities the symbols $\bar{x}^{\bar{\varkappa}}$, $\bar{g}_{\bar{\lambda}\bar{\mu}}$, etc. No ambiguity enters however in the subject matter of this chapter, if we merely write $\bar{\varphi}$ for the conjugate complex of a scalar φ.

2. Linear displacement. We can also assign in an X_n real and complex values to the original variables $\xi^\varkappa$. At every point we have a local E_n of the above mentioned type. It is then possible to introduce HERMITian tensorfields into the X_n which transform in this way:

$$g_{\lambda'\bar{\mu}'} = A^{\lambda\bar{\mu}}_{\lambda'\bar{\mu}'} g_{\lambda\bar{\mu}}, \qquad A^{\lambda}_{\lambda'} = \partial_{\lambda'} \xi^{\lambda}, \qquad A^{\bar{\mu}}_{\bar{\mu}'} = \partial_{\bar{\mu}'} \xi^{\bar{\mu}}.$$

A general linear displacement can be introduced as an expression of the form

$$\delta v^\varkappa = d v^\varkappa + \Gamma^\varkappa_{\mu\lambda} v^\lambda d\xi^\mu + \Gamma^\varkappa_{\mu\bar{\lambda}} v^{\bar{\lambda}} d\xi^\mu + \Gamma^\varkappa_{\bar{\mu}\lambda} v^\lambda d\xi^{\bar{\mu}} + \Gamma^\varkappa_{\bar{\mu}\bar{\lambda}} v^{\bar{\lambda}} d\xi^{\bar{\mu}},$$

$$\delta v^{\bar{\varkappa}} = d v^{\bar{\varkappa}} + \Gamma^{\bar{\varkappa}}_{\bar{\mu}\bar{\lambda}} v^{\bar{\lambda}} d\xi^{\bar{\mu}} + \Gamma^{\bar{\varkappa}}_{\bar{\mu}\lambda} v^{\lambda} d\xi^{\bar{\mu}} + \Gamma^{\bar{\varkappa}}_{\mu\bar{\lambda}} v^{\bar{\lambda}} d\xi^{\mu} + \Gamma^{\bar{\varkappa}}_{\mu\lambda} v^{\lambda} d\xi^{\mu},$$

where the Γ are $8n^3$ independent parameters, functions of $\xi^\varkappa$, $\xi^{\bar{\varkappa}}$. Their number can immediately be reduced to $4n^3$, when we assume that the covariant differentials of conjugate quantities are conjugate themselves. This makes $\Gamma^\varkappa_{\mu\bar{\lambda}}$ the conjugate of $\Gamma^{\bar{\varkappa}}_{\bar{\mu}\lambda}$, etc., an assumption already accounted for in our notation.

As in the case of the L_n we shall reduce the form of this displacement by special assumptions. To interpret them it is useful to map, for a special choice of coordinates, the X_n on an X_{2n} with real variables only, using the equation

$$\xi^\varkappa \overset{*}{=} \xi^{\varkappa_1} + i\xi^{\varkappa_2}, \qquad \xi^{\bar{\varkappa}} \overset{*}{=} \xi^{\varkappa_1} - i\xi^{\varkappa_2},$$

where the $\xi^{\varkappa_1}$ and the $\xi^{\varkappa_2}$ together form $2n$ real independent variables in the X_{2n}. In this X_{2n} the equations $\xi^\varkappa = \text{const}$, $\xi^{\bar{\varkappa}} = \text{const.}$ represent two families of $\infty^n X_n$ which, in analogy to the case where the X_{2n} is an R_2, may be called the *isotropic* X_n of the first and second kind.

The X_n of the first kind, $\xi^\varkappa = \text{const}$, and those of the second kind, $\xi^{\bar{\varkappa}} = \text{const}$, correspond to each other point by point through association of the points with conjugate complex coordinates. This implies a one to one corespondence of the linear elements $d\xi^{\bar{\varkappa}}$ at a point of $\xi^\varkappa = \text{const}$, to the linear elements $d\xi^\varkappa$ at the corresponding point of $\xi^{\bar{\varkappa}} = \text{const}$. We may call this *equipollence.*

The following assumptions concerning the Γ can now be interpreted geometrically.

a) $\Gamma^\varkappa_{\bar{\mu}\bar{\lambda}} = 0$, hence $\Gamma^{\bar{\varkappa}}_{\mu\lambda} = 0$. The n-direction of every isotropic X_n retains this property by parallel transport in a direction of this X_n. This may be expressed by calling the isotropic X_n *geodesic.*

b) $\Gamma^\varkappa_{\mu\bar{\lambda}} = 0$, hence $\Gamma^{\bar{\varkappa}}_{\bar{\mu}\lambda} = 0$. When the points of an X_n $\xi^\varkappa = \text{const}$ are moved along equipollent segments $(0, d\xi^{\bar{\varkappa}})$, the X_n passes into another X_n of the same kind; a similar property holds for the X_n of the other kind. We may express this by calling the isotropic X_n of each kind *parallel.*

c) $\Gamma^{\varkappa}_{\bar\mu\lambda} = 0$, hence $\Gamma^{\varkappa}_{\mu\bar\lambda} = 0$. Every vector in an isotropic X_n $\xi^{\varkappa} =$ const, if moved parallel in a direction contained in an isotropic X_n of the other kind to another X_n $\xi^{\varkappa} =$ const, will pass into an equipollent vector, and similarly for a vector in $\xi^{\bar\varkappa} =$ const. This expresses the equipollence of the isotropic X_n under this displacement. If b) and c) are satisfied there exist infinitesimal "parallelograms" of which two sides lie in an isotropic direction of the first kind and two in an isotropic direction of the second kind, the opposite sides being parallel in the sense of this displacement. Inside an isotropic X_n such infinitesimal parallelograms need not exist, for in this case the torsion must vanish, that is $\Gamma^{\varkappa}_{\lambda\mu} = \Gamma^{\varkappa}_{\mu\lambda}$, $\Gamma^{\bar\varkappa}_{\bar\lambda\bar\mu} = \Gamma^{\bar\varkappa}_{\bar\mu\bar\lambda}$ (see below).

If the conditions a), b), c) are satisfied, we have a displacement that can be formally expressed like one of the type L_n

$$\delta v^{\varkappa} = d v^{\varkappa} + \Gamma^{\varkappa}_{\mu\lambda} v^{\lambda} d\xi^{\mu}, \qquad \delta v^{\bar\varkappa} = d v^{\bar\varkappa} + \Gamma^{\bar\varkappa}_{\bar\mu\bar\lambda} v^{\bar\lambda} d\xi^{\bar\mu}.$$

We must, however, not forget that

$$d v^{\varkappa} = (\partial_{\mu} v^{\varkappa})\, d\xi^{\mu} + (\partial_{\bar\mu} v^{\varkappa})\, d\xi^{\bar\mu}; \qquad d v^{\bar\varkappa} = (\partial_{\mu} v^{\bar\varkappa})\, d\xi^{\mu} + (\partial_{\bar\mu} v^{\bar\varkappa})\, d\xi^{\bar\mu}.$$

Such a connection will be called a K_n.

3. Connection K_n. This connection is fully determined as far as quantities are concerned. For instance we have

$$\delta w_{\lambda} = d w_{\lambda} - \Gamma^{\varkappa}_{\mu\lambda} w_{\varkappa} d\xi^{\mu} \text{ (also its conjugate)},$$

$$\delta h_{\lambda\bar\mu} = d h_{\lambda\bar\mu} - \Gamma^{\varkappa}_{\nu\lambda} h_{\varkappa\bar\mu} d\xi^{\nu} - \Gamma^{\bar\varkappa}_{\bar\nu\bar\mu} h_{\lambda\bar\varkappa} d\xi^{\bar\nu} \text{ (also its conjugate)}.$$

In this displacement several covariant derivatives belong to one covariant differential. To $\delta v^{\varkappa}$, $\delta v^{\bar\varkappa}$ belong

$$\nabla_{\mu} v^{\varkappa} = \partial_{\mu} v^{\varkappa} + \Gamma^{\varkappa}_{\mu\lambda} v^{\lambda}; \qquad \nabla_{\bar\mu} v^{\bar\varkappa} = \partial_{\bar\mu} v^{\bar\varkappa} + \Gamma^{\bar\varkappa}_{\bar\mu\bar\lambda} v^{\bar\lambda},$$

$$\nabla_{\bar\mu} v^{\varkappa} = \partial_{\bar\mu} v^{\varkappa}; \qquad \nabla_{\mu} v^{\bar\varkappa} = \partial_{\mu} v^{\bar\varkappa}.$$

In the X_{2n} these equations can be written as one

$$\nabla_{\mathfrak{b}} v^{\mathfrak{c}} = \partial_{\mathfrak{b}} v^{\mathfrak{c}} + \Gamma^{\mathfrak{c}}_{\mathfrak{b}\mathfrak{a}} v^{\mathfrak{a}}, \qquad \mathfrak{a}, \mathfrak{b}, \mathfrak{c}, \dots = 1, \dots, n, \bar 1, \dots, \bar n.$$

As we deal with a K_n, some $\Gamma^{\mathfrak{c}}_{\mathfrak{b}\mathfrak{a}}$ vanish. The corresponding curvature tensor is

$$R_{\mathfrak{d}\mathfrak{c}\mathfrak{b}}{}^{\cdot\cdot\cdot\mathfrak{a}} = -2\partial_{[\mathfrak{d}} \Gamma^{\mathfrak{a}}_{\mathfrak{c}]\mathfrak{b}} - 2\Gamma^{\mathfrak{c}}_{[\mathfrak{d}|\mathfrak{e}|} \Gamma^{\mathfrak{e}}_{\mathfrak{c}]\mathfrak{b}}.$$

It has the following non-vanishing components:

$$R_{\nu\mu\lambda}{}^{\cdot\cdot\cdot\varkappa}, \quad R_{\bar\nu\bar\mu\bar\lambda}{}^{\cdot\cdot\cdot\bar\varkappa}, \quad R_{\bar\nu\mu\lambda}{}^{\cdot\cdot\cdot\varkappa}, \quad R_{\nu\bar\mu\bar\lambda}{}^{\cdot\cdot\cdot\bar\varkappa},$$

of which the last two sets satisfy the equations

$$R_{\bar\nu\mu\lambda}{}^{\cdot\cdot\cdot\varkappa} = -R_{\mu\bar\nu\lambda}{}^{\cdot\cdot\cdot\varkappa} = -\partial_{\bar\nu} \Gamma^{\varkappa}_{\mu\lambda}.$$

There is a torsion tensor

$$S_{\mu\lambda}{}^{\cdot\cdot\varkappa} = \Gamma^{\varkappa}_{[\mu\lambda]}; \qquad S_{\bar\mu\bar\lambda}{}^{\cdot\cdot\bar\varkappa} = \Gamma^{\bar\varkappa}_{[\bar\mu\bar\lambda]}.$$

There is also a second identity for the curvature tensor

$$R_{\bar{\nu}\mu\lambda}^{\cdot\cdot\cdot\varkappa} - R_{\bar{\nu}\lambda\mu}^{\cdot\cdot\cdot\varkappa} = 2\partial_{\bar{\nu}} S_{\mu\lambda}^{\cdot\cdot\varkappa} = 2\nabla_{\bar{\nu}} S_{\mu\lambda}^{\cdot\cdot\varkappa}, \text{ and its conjugate,}$$

together with

$$0 = 2\nabla_{[\mu} S_{\lambda\nu]}^{\cdot\cdot\varkappa} + 2S_{[\mu\lambda}^{\cdot\cdot\pi} S_{\nu]\pi}^{\cdot\cdot\varkappa},$$

and also identities of BIANCHI

$$\nabla_{[\mu} R_{\pi]\bar{\nu}\lambda}^{\cdot\cdot\cdot\varkappa} = S_{\mu\pi}^{\cdot\cdot\varrho} R_{\varrho\bar{\nu}\lambda}^{\cdot\cdot\cdot\varkappa}, \text{ (and its conjugate),}$$

$$\nabla_{[\mu} R_{\pi]\bar{\nu}\bar{\lambda}}^{\cdot\cdot\cdot\bar{\varkappa}} = S_{\mu\pi}^{\cdot\cdot\varrho} R_{\varrho\bar{\nu}\bar{\lambda}}^{\cdot\cdot\cdot\bar{\varkappa}}, \text{ (and its conjugate).}$$

A connection K_n can be made metrical by the introduction of a symmetrical HERMITian tensor $g_{\lambda\bar{\mu}} = g_{\bar{\mu}\lambda}$ satisfying the equation

$$\delta g_{\lambda\bar{\mu}} = 0.$$

A measurement can be introduced by $ds^2 = g_{\lambda\bar{\mu}} d\xi^\lambda d\xi^{\bar{\mu}}$. Such a connection is better called *unitary*, and is denoted by U_n. For a U_n we find, from $\delta g_{\lambda\bar{\mu}} = 0$

$$\Gamma_{\mu\lambda}^{\varkappa} = g^{\bar{\nu}\varkappa} \partial_\lambda g_{\mu\bar{\nu}}, \qquad \Gamma_{\bar{\mu}\bar{\lambda}}^{\bar{\varkappa}} = g^{\nu\bar{\varkappa}} \partial_{\bar{\lambda}} g_{\bar{\mu}\nu},$$

and therefore

$$R_{\nu\mu\lambda}^{\cdot\cdot\cdot\varkappa} = 0, \quad R_{\bar{\nu}\bar{\mu}\bar{\lambda}}^{\cdot\cdot\cdot\bar{\varkappa}} = 0.$$

There is now a third and fourth identity for the components of the curvature tensor that do not vanish, namely

$$R_{\bar{\nu}\mu\lambda\bar{\varkappa}} = -R_{\bar{\nu}\mu\bar{\varkappa}\lambda}; \qquad R_{\bar{\nu}\mu\lambda\bar{\varkappa}} - R_{\lambda\bar{\varkappa}\bar{\nu}\mu} = 2\nabla_\lambda S_{\bar{\nu}\bar{\varkappa}\mu} - 2\nabla_{\bar{\nu}} S_{\lambda\mu\bar{\varkappa}}.$$

Such a connection can be obtained, similarly to the affine connection with torsion (Ch. II), by introducing n independent vectors $\overset{i}{u}_\lambda$, $i = 1, 2, \ldots, n$ at each point, and building up a $g_{\lambda\bar{\mu}} = \sum \overset{i}{u}_\lambda \overset{i}{u}_{\bar{\mu}}$. As a rule the displacement does not carry these vector fields into themselves because of the non-vanishing of $R_{\bar{\nu}\mu\lambda}^{\cdot\cdot\cdot\varkappa}$. The $\overset{i}{u}_\lambda$ are called unitary vectors and satisfy the conditions

$$\overset{i}{u}_\lambda \underset{j}{u}^\lambda = \delta^i_j; \qquad \overset{i}{u}_{\bar{\lambda}} \underset{j}{u}^{\bar{\lambda}} = \delta^i_j,$$

if $\underset{i}{u}^\varkappa$ are the reciprocal vectors to the $\overset{i}{u}_\lambda$, so that a tensor $g^{\lambda\bar{\varkappa}}$ is uniquely defined as $g^{\lambda\bar{\varkappa}} = \sum \underset{i}{u}^\lambda \underset{i}{u}^{\bar{\varkappa}}$. The $\overset{i}{u}_\lambda$ can be considered as mutually orthogonal unit vectors. There exists a contracted curvature tensor $R_{\bar{\nu}\mu\lambda}^{\cdot\cdot\cdot\lambda}$ and a curvature scalar

$$R = R_{\bar{\nu}\mu\lambda}^{\cdot\cdot\cdot\lambda} g^{\bar{\nu}\mu}, \qquad \bar{R} = R_{\nu\bar{\mu}\lambda}^{\cdot\cdot\cdot\lambda} g^{\nu\bar{\mu}}.$$

It is also possible to define U_n of *constant curvature* $R_{\bar{\nu}\mu\lambda}^{\cdot\cdot\cdot\varkappa} = C A^\varkappa_{[\mu} g_{\bar{\nu}]\lambda}$, where C must be a constant on account of BIANCHI's identity. These U_n are applicable to the projective HERMITian geometries of FUBINI and CARTAN [1].

[1] SCHOUTEN and VAN DANTZIG: 1931 (17).

4. Analyticity. A complex function $\varphi = \alpha + i\beta$ of a complex variable $z = x + iy$ is called *analytical* when the RIEMANN-CAUCHY equations are satisfied. This can be expressed by the equation $\partial_{\bar{z}}\varphi = 0$, when $\partial_{\bar{z}} = \partial/\partial x - i\,\partial/\partial y$; or by the equivalent $\partial_z \bar{\varphi} = 0$, when $\partial_z = \partial/\partial x + i\,\partial/\partial y$. In the same way we call a function $\varphi(\xi^\varkappa)$ of the complex variables $\xi^\varkappa = \xi^{\varkappa_1} + i\,\xi^{\varkappa_2}$ analytical, if

$$\partial_{\bar{\mu}}\varphi = 0, \qquad \partial_{\bar{\mu}} = \partial_{\varkappa_1} - i\,\partial_{\varkappa_2}, \qquad \partial_{\varkappa_i} = \partial/\partial\,\xi^{\varkappa_i}$$

equivalent to

$$\partial_\mu \bar{\varphi} = 0, \qquad \partial_\mu = \partial_{\varkappa_1} + i\,\partial_{\varkappa_2}.$$

Quantities with analytical components are called analytical. This property is unchanged under coordinate transformations.

This property admits a simple geometrical interpretation in the X_{2n}. A scalar field is analytical in $\xi^\varkappa$ when it is constant in the X_n $\xi^{\bar{\varkappa}} = \text{const}$ and vice versa. The vector field v^c which is composed of the components $v^\varkappa$, $v^{\bar{\varkappa}}$ has, in the case of analyticity of $v^\varkappa$, $v^{\bar{\varkappa}}$ the property that its component in every isotropic X_n of one kind is equipollent to itself. An analytical transformation of the X_n corresponds to a transformation of the X_{2n} which carries the two families of isotropic X_n into themselves.

A displacement, which carries analytical quantities into analytical quantities is also called *analytical*; and so is the corresponding connection. Such a displacement

$$\delta v^\varkappa = d v^\varkappa + \Gamma^\varkappa_{\mu\lambda} v^\lambda d\xi^\mu, \qquad \delta v^{\bar{\varkappa}} = d v^{\bar{\varkappa}} + \Gamma^{\bar{\varkappa}}_{\bar{\mu}\bar{\lambda}} v^{\bar{\lambda}} d\xi^{\bar{\mu}}$$

must have for $dv^\varkappa$ the simple expressions, similar to those in L_n:

$$d v^\varkappa = (\partial_\mu v^\varkappa)\, d\xi^\mu, \qquad d v^{\bar{\varkappa}} = (\partial_{\bar{\mu}} v^{\bar{\varkappa}})\, d\xi^{\bar{\mu}},$$

and must further have the $\Gamma^\varkappa_{\mu\nu}$ analytical:

$$\partial_{\bar{\mu}} \Gamma^\varkappa_{\lambda\nu} = 0, \qquad \partial_\mu \Gamma^{\bar{\varkappa}}_{\bar{\lambda}\bar{\nu}} = 0.$$

This last equation is equivalent to

$$R_{\bar{\nu}\mu\lambda}{}^{\cdot\cdot\cdot\varkappa} = 0, \qquad R_{\nu\bar{\mu}\lambda}{}^{\cdot\cdot\cdot\varkappa} = 0.$$

This condition is therefore the necessary and sufficient condition for the analyticity of the connection K_n.

An analytical K_n has therefore a torsion tensor $S_{\mu\lambda}{}^{\cdot\cdot\varkappa}$, $S_{\bar{\mu}\bar{\lambda}}{}^{\cdot\cdot\bar{\varkappa}}$ and a curvature tensor $R_{\nu\mu\lambda}{}^{\cdot\cdot\cdot\varkappa}$, $R_{\bar{\nu}\bar{\mu}\bar{\lambda}}{}^{\cdot\cdot\cdot\bar{\varkappa}}$. For an analytical U_n the curvature tensor vanishes. Suppose, morover, that the torsion vanishes. Then we can show that the unitary vectors $\underset{i}{u}^\varkappa$ are now carried into themselves by parallel displacements. They must therefore be analytical, and gradient vectors of n analytical scalar fields $\overset{\varkappa}{x}$. These fields can be taken as coordinate variables, and form a CARTESian coordinate system. We have a *plane* HERMITian geometry in the U_n. With respect to this coordinate system we can write the ds^2 in the form $d\xi^1 d\xi^{\bar{1}} + \cdots$, or

$$ds^2 = \pm \sum d\xi^\varkappa d\xi^{\bar{\varkappa}}.$$

Such a plane HERMITian geometry has vanishing torsion, vanishing curvature tensor and an analytical connection[1].

5. Spin connections; introduction. In investigations connected with the spinning electron it has been shown that we can obtain a geometry of HERMITian quantities in an E_4 by taking as starting point a euclidean R_4 of the MINKOWSKI type. In such an R_4 there exists a hypersphere

$$g_{ij}x^i x^j = 1, \qquad i, j = 0, 1, 2, 3, \qquad -g_{00} = g_{11} = g_{22} = g_{33} = +1,$$
$$g_{ij} = 0, \qquad i \neq j$$

of signature $-+++$. The $\overline{\infty}^3$ straight lines of this hypersphere determine the directions of ∞^4 vectors. These vectors can be represented by the ∞^4 points of an auxiliary E_4, the so-called *spin-space*. The orthogonal transformations which carry this hypersphere into itself (LORENTZ transformations) can be used, as we will show, to define HERMITian quantities in this E_4, especially a HERMITian tensor. If now a V_4, locally of the MINKOWSKI type, is given, then the problem arises of defining a connection with a displacement which allows comparison of these spin-spaces. This displacement must, therefore, map the vectors in the straight lines of the hypersphere in the local R_4 at one point onto the corresponding vectors of an adjacent point. This connection is of importance for the DIRAC theory of the spinning electron.

6. Spinspace[2]. Between the ∞^6 points r^c, $c = 0, 1, 2, \ldots, 5$ of a euclidean R_6 and the ∞^6 bivectors r^{AC}, $A, B, C, \ldots = 1, \ldots 4$ of an affine E_4 a one to one correspondence can be established

$$r^{AC} = r^c \chi_c^{\cdot AC}.$$

The $\chi_c^{\cdot AC}$ themselves can be taken as a tensor with its lower index c in R_6 and its indices AC in E_4. To the fundamental tensor g^{ij} corresponds $\frac{1}{2}\varepsilon^{[ABCD]}$, where $\varepsilon^{[ABCD]}$ is the unit four vector of E_4. The corresponding relation between contravariant and covariant quantities in R_6 and E_4 follow from the corresponding equations

$$r_a = g_{ab} r^b, \quad r^b = g^{ab} r_a \rightarrow r_{AB} = \tfrac{1}{2}\varepsilon_{ABCD} r^{CD}, \quad r^{CD} = \tfrac{1}{2}\varepsilon^{ABCD} r_{AB}.$$

We have $r^{AB} r_{BC} = C \alpha_C^A$, where C is an invariant and α_C^A is the unit tensor of E_4.

To orthogonal vectors in R_6 belong bivectors in *involution* of E_4,

$$r^c s_c = 0 \rightarrow r^{[AB} s^{CD]} = 0.$$

There exist in E_4 six bivectors in involution corresponding to six orthogonal unit vectors $\underset{a}{i^c}$ of R_6. This imposes the following conditions upon $\chi_c^{\cdot AC}$

$$\chi_{(a}^{\cdot AC} \chi_{b)BC} = g_{ab} \alpha_B^A.$$

---

[1] Comp. KÄHLER: 1932 (13).

[2] SCHOUTEN: 1933 (2). — SCHOUTEN-VAN DANTZIG: 1933 (6); VEBLEN: 1933 (9), 1933 (10).

To all ∞^2 bivectors r_{AB} satisfying the equation $v^A r_{AB} = 0$, where v^A is a given point of E_4, belong the ∞^2 points of an R_2 in the R_6. From $v^A r_{AB} = 0$ follows $r_{[AB} r_{CD]} = 0$, which corresponds to $r_c r^c = 0$, the zero sphere in R_6. We see therefore that to a point v^A of E_4 corresponds a simple bivector in a plane R_5 in the zero sphere of R_6.

We now single out an R_5 in R_6 by the condition that $\underset{5}{i}^c$ be a fixed vector. This corresponds to the fixing of a bivector χ_{AB} in E_4. The points of E_4 correspond now to the R_2 in the zero cone of this R_5.

There are five mixed tensors determined by this choice

$$\alpha_{a\cdot A}^{\cdot C} = -\chi_a^{\cdot CB}\chi_{BA} = \chi^{CD}\chi_{aDA}, \quad \text{in short} \quad \alpha = -\chi_a\chi = \chi\chi_a,$$

which satisfy the conditions

$$\alpha_{(ab)\cdot A}^{\cdot\cdot C} = \alpha_{(a\cdot|B|}^{\cdot C}\alpha_{b)\cdot A}^{\cdot B} = -\chi_{(a}^{\cdot CB}\chi_{b)BA} = g_{ab}\alpha_A^C,$$

or more briefly

$$\alpha_{(ab)} = \alpha_{(a}\alpha_{b)} = g_{ab}.$$

Now we take the R_6 of signature $---+++$,

$$s^2 = -(r^0)^2 + (r^1)^2 + (r^2)^2 + (r^3)^2 - (r^4)^2 - (r^5)^2.$$

In this case we have

$$-\alpha_0\alpha_0 = \alpha_1\alpha_1 = \alpha_2\alpha_2 = \alpha_3\alpha_3 = -\alpha_4\alpha_4 = +1,$$
$$\alpha_i\alpha_j = -\alpha_j\alpha_i = 0, \qquad i \neq j, \qquad\qquad i, j = 0, 1, \ldots, 4.$$

The α_a, which may all be taken real, behave like the units of a *sedenion* system, that is, a hypercomplex number system which can be built up by linear combination of 16 units 1; $\alpha_0, \alpha_1, \alpha_2, \alpha_3$; $\alpha_{ab} = \alpha_a\alpha_b$, $\alpha_{abc} = \alpha_a\alpha_b\alpha_c$, $\alpha_4 = \alpha_0\alpha_1\alpha_2\alpha_3$, and satisfying the associative law

$$\alpha_a(\alpha_b\alpha_c) = (\alpha_a\alpha_b)\alpha_c = \alpha_a\alpha_b\alpha_c.$$

The zero cone of R_5 has the equation

$$0 = -(r^0)^2 + (r^1)^2 + (r^2)^2 + (r^3)^2 - (r^4)^2.$$

If we now introduce the coordinates

$$\varrho^1 = \frac{r^1}{r^0}, \ldots, \varrho^4 = \frac{r^4}{r^0}$$

this zero cone passes into the fundamental sphere

$$(\varrho^1)^2 + (\varrho^2)^2 + (\varrho^3)^2 - (\varrho^4)^2 = 1$$

of a Minkowski R_4, of which the vectors in the generating straight lines correspond to the points of the E_4. As quantities expressing this correspondence the sedenions α can be taken. These sedenions can be expressed in every coordinate system of E_4 as matrices. Such a matrix is a spinmatrix of quantum theory. The E_4 is therefore a *spinspace*, its quantities are *spinquantities*.

7. HERMITian quantities in spinspace. The sedenion tensors $\alpha_0, \alpha_1, \alpha_2, \alpha_3$ are determined but for LORENTZ transformations. The $\alpha_4 = \alpha_0 \alpha_1 \alpha_2 \alpha_3 = \alpha_{[0} \alpha_1 \alpha_2 \alpha_{3]}$ remains however invariant (but for the sign). The $\alpha_{4\cdot A}^{\cdot C}$ determines a matrix of which the elementary divisors are $\lambda - i$, $\lambda - i$, $\lambda + i$, $\lambda + i$ and can therefore be written with the aid of 4 contravariant measuring spinvectors $\underset{a}{e}^C$ and the corresponding covariant vectors $\overset{a}{e}_A \left(\alpha_A^C = \underset{a}{e}^C \overset{a}{e}_A\right)$ as follows

$$\alpha_{4\cdot A}^{\cdot C} = -i \underset{1}{e}^C \overset{1}{e}_A - i \underset{2}{e}^C \overset{2}{e}_A + i \underset{3}{e}^C \overset{3}{e}_A + i \underset{4}{e}^C \overset{4}{e}_A. \quad (i = \sqrt{-1})$$

This shows that there exist, in E_4, two invariant E_2, the E_2 of $\underset{1}{e}^C$, $\underset{2}{e}^C$ and the E_2 of $\underset{3}{e}^C$, $\underset{4}{e}^C$, which have only the origin in common. We call these planes E_2 and $\overline{E}_2$.

Every vector v^C in spinspace can be decomposed in the E_2 and the $\overline{E}_2$. We can write this in the form

$$v^C = \underset{0}{i}{}^C_{\cdot A} v^A + \underset{0}{\bar{i}}{}^C_{\cdot A} v^A,$$

where $\underset{0}{i}$, $\underset{0}{\bar{i}}$ are the unit tensors in the E_2, $\overline{E}_2$,

$$\underset{0}{i} = \tfrac{1}{2}(1 + i\alpha_4), \qquad \underset{0}{\bar{i}} = \tfrac{1}{2}(1 - i\alpha_4)$$

which is the abbreviation of

$$\underset{0}{i}{}^C_{\cdot A} = \underset{1}{e}^C \overset{1}{e}_A + \underset{2}{e}^C \overset{2}{e}_A = \tfrac{1}{2}(\alpha_A^C + i\alpha_{4\cdot A}^{\cdot C}), \quad \text{and similarly for } \underset{0}{\bar{i}}.$$

The tensors belonging to α_0, α_1, α_2, α_3 can now be decomposed

$$\alpha_\varkappa = \beta_\varkappa + \bar{\beta}_\varkappa, \qquad (\varkappa = 0, 1, 2, 3)$$

where $\beta_\varkappa$ is a tensor which belongs with the contravariant index to E_2 and with the covariant index to $\overline{E}_2$, and the $\bar{\beta}_\varkappa$ behaves in the opposite way. The other two parts of $\alpha_\varkappa$, belonging to E_2 and $\overline{E}_2$ alone, vanish. We express $\beta_\varkappa$, $\bar{\beta}_\varkappa$ as follows:

$$\beta_\varkappa = \beta_{\varkappa\cdot\mathfrak{A}}^{\cdot\mathfrak{c}}, \qquad \bar{\beta}_\varkappa = \bar{\beta}_{\varkappa\cdot\mathfrak{a}}^{\cdot\mathfrak{C}}$$

where the indices $\mathfrak{a}, \mathfrak{b}, \mathfrak{c}, \ldots$ belong to E_2, $\mathfrak{A}, \mathfrak{B}, \mathfrak{C}, \ldots$ to $\overline{E}_2$. We have for the β the property, reminiscent of the sedenion properties

$$\beta_{(i}\bar{\beta}_{j)} = \underset{0}{i}\, g_{ij}, \quad \bar{\beta}_{(i}\beta_{j)} = \underset{0}{\bar{i}}\, g_{ij}, \quad -g_{00} = g_{11} = g_{22} = g_{33} = +1, \quad g_{ij} = 0, \quad i \neq j.$$

These two formulas are the inverse of each other for $k = 1, 2, 3$. For $k = 0$, there is a change of sign. For every LORENTZ transformation these formulas take another form.

The $\underset{0}{i}$ can be completed to a set of quaternions, as can the $\underset{0}{\bar{i}}$, e. g. $i_1 = i\beta_0\bar{\beta}_1 = -\beta_2\bar{\beta}_3$, etc. It can now be shown that a choice of the coordinate systems in E_2, $\overline{E}_2$ may be made so that the i and $\bar{i}$, and hence also the β and $\bar{\beta}$, obtain conjugate complex components. This is possible only because we started with a MINKOWSKI form of linear element in R_4. We have to restrict the coordinate transformations in E_2, $\bar{E}_2$ to linear homogeneous ones with complex conjugate coefficients, in order to make the choice invariant. The determinants of the β satisfy, under these transformations, the equations

$$|\beta_0| = -|\beta_\varkappa|, \qquad |\bar{\beta}_0| = -|\bar{\beta}_\varkappa|. \qquad (\varkappa = 1, 2, 3)$$

We postulate, secondly, that the transformations will have real determinants, that is,

$$|\beta_0| = -|\beta_1| = -|\beta_2| = -|\beta_3| = +1,$$
$$|\bar{\beta}_0| = -|\bar{\beta}_1| = -|\bar{\beta}_2| = -|\bar{\beta}_3| = +1.$$

Now we have established in E_4 a system of HERMITian quantities with a group of transformations under which they remain HERMITian. It can even be shown that through these assumptions symmetrical HERMITian densities of order -1 are determined[1].

8. Spin connections. We consider a V_4 which has, at each point, a local R_4 of MINKOWSKI character. To every R_4 is associated a local spin space determined by the straight lines on the fundamental quadric. It is possible to define an indefinite number of linear connections which map these local spin spaces upon each other. As the β, $\bar{\beta}$ have the character of units, we may postulate that their covariant differentials vanish. Nevertheless it is not in general possible to define the spin-connection uniquely by means of the quantities in the V_4. There is however one exception. Covariant differentiation of contravariant and covariant spinvector-densities of weight $+\frac{1}{2}$ and $-\frac{1}{2}$, respectively is uniquely determined.

To show this, let the displacement of a spinvector be

$$\nabla_j v^c = \partial_j v^c + \Lambda'^{c}_{ja} v^a. \qquad j \text{ refers to } V_4$$

The $\beta^{k\mathfrak{c}}_{\cdot\cdot\mathfrak{A}}$ is a quantity with the k in V_4 and with the $\mathfrak{c}$ and $\mathfrak{A}$ in E_2 and $\bar{E}_2$. Therefore we can write under assumptions similar to those of art. 2, taking a nonholonomic system of reference in the V_4,

$$\nabla_j \beta^{k\mathfrak{c}}_{\cdot\cdot\mathfrak{A}} = \partial_j \beta^{k\mathfrak{c}}_{\cdot\cdot\mathfrak{A}} + \Gamma^k_{ji} \beta^{i\mathfrak{c}}_{\cdot\cdot\mathfrak{A}} + \Lambda'^{\mathfrak{c}}_{j\mathfrak{C}} \beta^{k\mathfrak{C}}_{\cdot\cdot\mathfrak{A}} - \Lambda'^{\mathfrak{C}}_{j\mathfrak{A}} \beta^{k\mathfrak{c}}_{\cdot\cdot\mathfrak{C}}.$$

When $\nabla_j \beta^{k\mathfrak{c}}_{\cdot\cdot\mathfrak{A}}$ vanishes, we have, as $\partial_j \beta^{k\mathfrak{c}}_{\cdot\cdot\mathfrak{A}} = 0$,

$$2\Gamma^{ik}_j = -\Lambda'^{\mathfrak{c}}_{j\mathfrak{C}} \beta^{k\mathfrak{C}}_{\cdot\cdot\mathfrak{A}} \bar{\beta}^{i\mathfrak{A}}_{\cdot\cdot\mathfrak{c}} + \Lambda'^{\mathfrak{C}}_{\mathfrak{A}j} \bar{\beta}^{i\mathfrak{A}}_{\cdot\cdot\mathfrak{c}} \beta^{k\mathfrak{c}}_{\cdot\cdot\mathfrak{C}}.$$

As $\Gamma^{ik}_i = 0$, we have

$$0 = -\Lambda'^{\mathfrak{a}}_{j\mathfrak{a}} + \Lambda'^{\mathfrak{A}}_{j\mathfrak{A}}, \quad (\text{imaginary part of } \Lambda'^{\mathfrak{a}}_{j\mathfrak{a}} = 0)$$

---

[1] SCHOUTEN: 1931 (18).

but we cannot express the Λ in terms of Γ. If now we write $\Lambda_{j\mathfrak{a}}^{\mathfrak{c}}$, $\Lambda_{j\mathfrak{A}}^{\mathfrak{C}}$ for the parameters of covariant differentiation of contravariant and covariant spinvector densities of weight $+\frac{1}{2}$ and $-\frac{1}{2}$, respectively,

$$\Lambda_{j\mathfrak{a}}^{\mathfrak{c}} = \Lambda'^{\mathfrak{c}}_{j\mathfrak{a}} - \tfrac{1}{2}\underset{0}{i}{}^{\cdot\mathfrak{c}}_{\mathfrak{a}}\overset{r}{\Lambda}{}^{b}_{jb}; \qquad \Lambda_{j\mathfrak{A}}^{\mathfrak{C}} = \Lambda'^{\mathfrak{C}}_{j\mathfrak{A}} - \tfrac{1}{2}\underset{0}{i}{}^{\cdot\mathfrak{C}}_{\mathfrak{A}}\overset{r}{\Lambda}{}^{\mathfrak{B}}_{j\mathfrak{B}}$$

$\left(\overset{r}{\Lambda}{}^{\mathfrak{b}}_{j\mathfrak{b}}\right.$ is the real part of $\left.\Lambda^{\mathfrak{b}}_{j\mathfrak{b}}\right)$; then we can solve the equations for Λ and get, if $\beta^i\beta^k = \beta^{ik}$

$$\Lambda_{j\mathfrak{a}}^{\mathfrak{c}} = -\tfrac{1}{4}\Gamma_{ikj}\beta^{ik\mathfrak{c}}_{\cdot\cdot\cdot\mathfrak{a}}, \qquad \Lambda_{j\mathfrak{A}}^{\mathfrak{C}} = -\tfrac{1}{4}\Gamma_{ikj}\bar{\beta}^{ik\mathfrak{C}}_{\cdot\cdot\cdot\mathfrak{A}}.$$

If we had taken a weight different from $+\frac{1}{2}$ or $-\frac{1}{2}$, we should not have had a unique solution. For the physical application it is sufficient that at any rate one type of vector density allows unique determination of the connection by the Γ of the V_4.

9. Remarks. The spinquantities of the previous articles appeared first as matrices in Dirac's theory of the spinning electron. As long as we have a Minkowski space (special relativity) this means that we study tensors with the aid of a preferred coordinate system. In a V_4 there are ∞^4 local spin spaces, and it is necessary to introduce also the transformation schemes. In Minkowski space the spinquantities appeared first, after a suggestion by Ehrenfest, as so-called spinors[1], of which the analysis has been given by van der Waerden, Schouten, Laporte and Uhlenbeck[2]. The relations to sedenions were given in full detail by Schouten[2], who also showed the possibility of a spin-connection. The relation of the spinvectors to the straight lines of the fundamental quadric in Minkowski space, which removed all artificiality from spinspace, was indicated by Veblen and fully constructed by Schouten and by Veblen. Schouten also showed the way in which spinquantities enter into a five-dimensional theory[2], and together with van Dantzig a related theory of projective connections[3] (Ch. V).

Chapter V.

Projective connections.

1. Introduction. We have seen, in Chapter III, that the paths of an A_n are not changed by a projective transformation

(1.1) $$'\Gamma^{\varkappa}_{\mu\lambda} = \Gamma^{\varkappa}_{\mu\lambda} + 2p_{(\mu}A^{\varkappa}_{\lambda)}. \qquad p_\lambda = \text{arbitrary vector}$$

The problem arose of associating with this group of transformations a single "projective" connection, which will take the place of the infinite

[1] See van der Waerden: 1929 (17).

[2] Schouten: 1931 (18). — Laporte-Uhlenbeck: 1931 (31); Veblen: 1933 (9), 1933 (10).

[3] Schouten and van Dantzig: 1932 (3) — 1933 (6).

number of L_n connections. The introduction of the parameters $\Pi^{\varkappa}_{\mu\lambda}$ was one step, but it was not yet sufficient, because they depend on a special choice of coordinates. We must try to continue in the same direction of research, changing the group of transformations. This has been done and has lead to important results. It seems however more useful to attack the problem from another side and use as the starting point the fundamental principle of differential geometry, as formulated in Ch. I. We take an X_n and associate with every point a local projective space D_n. We ask for the linear connections associable with this configuration. It can then be shown that an infinite number of L_n related by (1.1) can be obtained from this *projective connection.*

There is a fundamental principle involved in this independent construction of a projective connection. Historically, the L_n came first; there exists therefore a certain tendency to relate connections to L_n. This is similar to the way plane projective, affine and conformal geometries were developed. First these geometries were studied as aspects of euclidean geometry, the oldest. Later, however, it was recognized that each of these geometries could be independently established, and taken as center of reference for the other geometries. Projective geometry was first the study of those properties of euclidean geometry which are invariant under projective transformation. Later it was recognized that euclidean geometry was that branch of projective geometry in which certain absolute elements are invariant. A similar process is now being undertaken in the theory of displacements. At present the independent construction of projective connections is well established, and the independent construction of other connections is well under way.

A text-book dealing with the subject matter of this chapter is VEBLEN's "Projective relativity". We follow here the independent construction of projective differential geometry due to VAN DANTZIG[1]. A related theory of conformal connections is due to CARTAN[2].

2. X_n with local D_n. We introduce into the X_n homogeneous coordinates $x^0, x^1, x^2, x^3, \ldots, x^n$, in short $x^\varkappa$, $\varkappa = 0, 1, \ldots, n$. All systems $y^\varkappa = \lambda x^\varkappa$ determine the same point. We subject these coordinates to the group $\mathfrak{H}_{n+1}$ of transformations

$$(\mathfrak{H}_{n+1}) \qquad x^{\varkappa'} = f^{\varkappa'}(x^\varkappa)$$

restricting the f to homogeneous functions of the first degree in the x (not necessarily linear). This is essentially the old $\mathfrak{G}_n$. Apart from this coordinate transformation we also allow the point transformation

$$(F) \qquad \bar{x}^\varkappa = \varrho x^\varkappa,$$

[1] VAN DANTZIG: 1932 (1).

[2] CARTAN: 1923 (2). — SCHOUTEN: 1924 (10); 1926 (1) — see also Ch. III, art. 4.

where ϱ is a function of $x^\varkappa$ of degree zero. The group (F) does not change the points of X_n. This X_n we call H_n. Points in H_n are therefore not identical with points in X_n. We may indicate this by calling a *spot* the series of ∞^1 points of H_n corresponding with one point of X_n.

Furthermore we admit only functions $f(x^\varkappa)$, homogeneous of degree $\mathfrak{r}$ in $x^\varkappa$, and therefore satisfying the condition

$$f(\varrho x^\varkappa) = \varrho^{\mathfrak{r}} f(x^\varkappa),$$

equivalent to the EULER equation

$$x^\mu \partial_\mu f = \mathfrak{r} f, \qquad \partial_\mu = \partial/\partial x^\mu.$$

At every point P of the X_n we can define a local projective space D_n. With the aid of

$$A^{\varkappa'}_{\varkappa} = \partial_\varkappa x^{\varkappa'},$$

$$A^{\varkappa}_{\varkappa'} = \partial_{\varkappa'} x^{\varkappa},$$

we can here define a point calculus (Ch. I, art. 4), in which the $A^{\varkappa'}_{\varkappa}$ have the function of the $\mathfrak{A}^{c'}_{c}$. Here we have to discriminate between the transformations $(\mathfrak{H}_{n+1})$ and (F). A tensor of degree $\mathfrak{r}$ transforms under $(\mathfrak{H}_{n+1})$ as in this example

$$v_{\lambda'\mu'}^{\cdot\;\cdot\;\varkappa'} = A^{\lambda\mu\varkappa'}_{\lambda'\mu'\varkappa} v_{\lambda\mu}^{\cdot\;\cdot\;\varkappa}$$

and under (F) as follows

$$\bar{v}_{\lambda\mu}^{\cdot\;\cdot\;\varkappa} = \varrho^{\mathfrak{r}} v_{\lambda\mu}^{\cdot\;\cdot\;\varkappa}.$$

A tensor $v_{\lambda_1 \ldots \lambda_s}^{\cdot\;\cdot\;\cdot\;\varkappa_1 \ldots \varkappa_t}$ of degree $\mathfrak{r}$ has an invariant, the *excess* (or weight) $\varepsilon = \mathfrak{r} + s - t$.[1] We study only tensors of excess zero.

This space D_n, which so far has been defined independently of the $x^\varkappa$ with the exception of the $A^{\varkappa'}_{\varkappa}$, $A^{\varkappa}_{\varkappa'}$, can be more closely related to the H_n by the following property which has no analogue in L_n. The point $x^\varkappa$ of H_n is itself a vector of degree 1 in D_n. Indeed, as a result of EULER's equation

$$x^\mu \partial_\mu x^{\varkappa'} = x^\mu A^{\varkappa'}_\mu = x^{\varkappa'}.$$

This point $x^\varkappa$ can be taken as "point of contact" of the D_n.

Another difference with L_n is that in this case $dx^\varkappa$ is a vector only with respect to $(\mathfrak{H}_{n+1})$, but not with respect to (F):

$$d\bar{x}^\varkappa = \varrho(dx^\varkappa + x^\varkappa d\ln\varrho).$$

There is, therefore, in general no point in D_n corresponding to $dx^\varkappa$.

3. Projective derivative. This behavior of the $dx^\varkappa$ makes it in general impossible, to define a covariant differential. But we can define a projective derivative[2]

$$\nabla_\mu v^\varkappa = \partial_\mu v^\varkappa + \Pi^\varkappa_{\mu\lambda} v^\lambda + \varepsilon Q_\mu v^\varkappa,$$

$$\nabla_\mu w_\lambda = \partial_\mu w_\lambda - \Pi^\varkappa_{\mu\lambda} w_\varkappa + \varepsilon Q_\mu w_\lambda,$$

[1] VEBLEN: 1929 (28).

[2] The Π and Q used in this chapter have a meaning different from the Π and Q used in Ch. III and II.

which follows from assumptions parallel to those of Ch. II, art. 2. The $\Pi_{\lambda\mu}^{\varkappa}$ form a system of $(n+1)^3$ functions of the $x^\varkappa$ of degree -1, and the Q_μ form a system of $n+1$ functions of the same degree. Hence

$$x^\nu \partial_\nu \Pi_{\mu\lambda}^{\varkappa} = -\Pi_{\mu\lambda}^{\varkappa},$$
$$x^\nu \partial_\nu Q_\mu = -Q_\mu .$$

The Π transform in the ordinary way under transformations of the $x^\varkappa$

$$\Pi_{\mu'\lambda'}^{\varkappa'} = A_{\mu'\lambda'\varkappa}^{\mu\lambda\varkappa'} \Pi_{\mu\lambda}^{\varkappa} + A_\varkappa^{\varkappa'} \partial_{\mu'} A_{\lambda'}^{\varkappa}.$$

There is a torsion tensor, homogeneous of degree -1,

$$S_{\mu\lambda}^{\cdot\cdot\varkappa} = \Pi_{[\mu\lambda]}^{\varkappa},$$

but also two new tensors (obtained from $\nabla_\mu x^\varkappa$)

$$P_{\cdot\lambda}^{\varkappa} = \Pi_{\mu\lambda}^{\varkappa} x^\mu ,$$
$$Q_{\cdot\mu}^{\varkappa} = \Pi_{\mu\lambda}^{\varkappa} x^\lambda = P_{\cdot\mu}^{\varkappa} + 2 S_{\mu\lambda}^{\cdot\cdot\varkappa} x^\lambda .$$

As $x^\mu \nabla_\mu v^\varkappa = P_{\cdot\mu}^{\varkappa} v^\mu$, $x^\mu \nabla_\mu w_\lambda = -P_{\cdot\lambda}^{\mu} w_\mu$, we see that the operator $x^\mu \nabla_\mu$ defines a projective transformation in D_n for vectors, which depends on points of H_n, because $x^\mu \nabla_\mu \lambda v^\varkappa = \lambda x^\mu \nabla_\mu v^\varkappa$, when λ is of degree 0.

Covariant differentiation of tensors of higher order follows in the usual way (Ch. II, art. 2).

There is a curvature tensor, homogeneous of degree -2

$$N_{\nu\mu\lambda}^{\cdot\cdot\cdot\varkappa} = -2\partial_{[\nu} \Pi_{\mu]\lambda}^{\varkappa} - 2\Pi_{[\nu|\pi|}^{\varkappa} \Pi_{\mu]\lambda}^{\pi} .$$

It satisfies the identities (I) $N_{\nu\mu\lambda}^{\cdot\cdot\cdot\varkappa} = -N_{\mu\nu\lambda}^{\cdot\cdot\cdot\varkappa}$ and

(II) $$N_{[\nu\mu\lambda]}^{\cdot\cdot\cdot\varkappa} = 4 S_{[\nu\lambda}^{\cdot\cdot\pi} S_{\mu]\pi}^{\cdot\cdot\varkappa} + 2\nabla_{[\nu} S_{\mu\lambda]}^{\cdot\cdot\varkappa} + 2 Q_{[\nu} S_{\mu\lambda]}^{\cdot\cdot\varkappa}$$

and Bianchi's identity

(III) $$\nabla_{[\pi} N_{\nu\mu]\lambda}^{\cdot\cdot\cdot\varkappa} = -2 S_{[\nu\mu}^{\cdot\cdot\varrho} N_{\pi]\varrho\lambda}^{\cdot\cdot\cdot\varkappa} - 2 A_{[\nu}^{\varrho} Q_\mu N_{\pi]\varrho\lambda}^{\cdot\cdot\cdot\varkappa} .$$

For the symbol $\nabla_{[\varkappa} \nabla_{\mu]}$ we get terms that do not occur in L_n:

$$\nabla_{[\nu} \nabla_{\mu]} v^\varkappa = -\tfrac{1}{2} N_{\nu\mu\lambda}^{\cdot\cdot\cdot\varkappa} v^\lambda + S_{\nu\mu}^{\cdot\cdot\varrho} \nabla_\varrho v^\varkappa + A_{[\nu}^{\pi} Q_{\mu]} \nabla_\pi v^\varkappa + \mathfrak{r}\, U_{\nu\mu} v^\varkappa ,$$
$$\nabla_{[\nu} \nabla_{\mu]} w_\lambda = +\tfrac{1}{2} N_{\nu\mu\lambda}^{\cdot\cdot\cdot\varkappa} w_\varkappa + S_{\nu\mu}^{\cdot\cdot\varrho} \nabla_\varrho w + A_{[\nu}^{\pi} Q_{\mu]} \nabla_\pi w_\lambda + \mathfrak{r}\, U_{\nu\mu} w_\lambda ,$$

where

$$U_{\nu\mu} = \nabla_{[\nu} Q_{\mu]} - S_{\nu\mu}^{\cdot\cdot\lambda} Q_\lambda - A_{[\nu}^{\lambda} Q_{\mu]} Q_\lambda .$$

4. Projective differential. If we define

$$\delta v^\varkappa = d x^\mu \nabla_\mu v^\varkappa ,$$

then the $\delta v^\varkappa$ transform under $(\mathfrak{H}_{n+1})$ like the components of a vector, but not under (F), because

$${}'\delta v^\varkappa = \varrho^{\mathfrak{r}} \{\delta v^\varkappa + (P_{\lambda}^{\cdot\varkappa} v^\lambda + \mathfrak{r} Q v^\varkappa)\, d \ln \varrho\},$$
$$Q = 1 + \lambda^\mu Q_\mu .$$

Now we impose upon the parameters Π the conditions

$$P_\lambda^{\cdot\varkappa} = P A_\lambda^\varkappa; \qquad Q = 0.$$

In this case we have a covariant differential, and therefore the possibility of mapping consecutive D_n upon each other by taking the covariant differential zero. A projective connection is thus defined, and consecutive D_n are mapped projectively on each other. In this representation the $\Pi_{\mu\lambda}^{\varkappa}$ are determined but for multiples of $A_\lambda^\varkappa$ and the Q_μ remain arbitrary, subject only to the given restrictions. The point of contact of a D_n does not, in general, remain a point of contact during a parallel displacement.

5. Relation to a metrical geometry[1]. The transition to a metrical geometry can be performed independently of any previous transition to an A_n by introducing immediately into the D_n of the projective connection H_n a symmetrical tensor $G_{\lambda\mu}$. This can be interpreted as a quadric on which lie the points $v^\varkappa$ for which $G_{\lambda\mu} v^\lambda v^\mu = 0$. If we now postulate that the point of contact $x^\varkappa$ does not lie on the quadric, we may write

$$G_{\lambda\mu} x^\lambda x^\mu = \omega^2,$$

and normalize the $G_{\lambda\mu}$ by giving the ω^2 as a fixed number. Now a euclidean metric can be constructed in the D_n. To a point $v^\varkappa$ can be assigned a modulus $\sqrt{G_{\lambda\mu} v^\lambda v^\mu}$. As the modulus of $x^\varkappa$ is ω we can introduce a unit point of contact $q^\varkappa = \omega^{-1} x^\varkappa$, which can be defined as center of the quadric. The $D_{n-1}\, q_\mu = G_{\lambda\mu} x^\lambda$ can then be taken as the D_{n-1} at infinity.

Functions ξ^k, $k = 1, 2, \ldots, n$, of degree zero in the $x^\varkappa$, independent and satisfying

$$q^\mu \partial_\mu \xi^k = 0$$

can now be taken as non-homogeneous coordinates; the ξ^k transform according to the group $\mathfrak{G}_n$. There is a unit tensor $A_\nu^k = \partial_\nu \xi^k$, $q^\nu A_\nu^k = 0$, and a unit tensor A_i^k. With respect to $\mathfrak{G}_n$ ordinary affine tensors can be defined

$$v_{i'}^{\cdot k'} = A_{i' k}^{l k'} v_l^{\cdot k}.$$

To a point in D_n a vector in this metrical space is now uniquely determined. We write, identifying the vectors v^k with the contravariant points v^ν in the $E_{n-1}\, q_\nu = 0$: $v^k = A_\nu^k v^\nu$, if $v^\nu q_\nu = 0$, and also $w_i = A_i^\lambda w_\lambda$, if $w_\lambda q^\lambda = 0$, and $A_k^\varkappa$ is defined by $A_\lambda^k A_i^\lambda = A_i^k$, $q_\nu A_k^\nu = 0$.

Every point can be written as a sum of a vector and a multiple of $q^\varkappa$:

$$v^\varkappa = {}'v^\varkappa + v q^\varkappa, \quad \text{where} \quad {}'v^\varkappa = A_\lambda^\varkappa v^\lambda + q_\lambda q^\varkappa v^\lambda. \qquad v = -v^\nu q_\nu$$

In the same way we can associate to every projective tensor an affine tensor which is identified with those projective tensors that admit, with respect to every index, inner multiplications with

$$\overline{A}_\lambda^\nu = A_\lambda^\nu + q_\lambda q^\nu.$$

[1] Schouten-van Dantzig: 1932 (4) — 1933 (6).

As $q^\lambda \partial_{[\lambda} q_{\mu]} = 0$ we find that $\partial_{[\lambda} q_{\mu]}$ itself is an affine tensor, denoting a null system in the D_{n-1} at infinity $q_\lambda = 0$, a fact of importance for projective relativity.

In this way an affine connection can be constructed in the original projective connection, taking at every point a fixed covariant vector q_λ.[1]

The symmetrical tensor

$$g_{\lambda\mu} = G_{\lambda\mu} + q_\lambda q_\mu$$

can then be taken as a RIEMANNian fundamental tensor.

If we take the covariant derivative of an affine tensor, then this covariant derivative has a part which is an affine tensor. In this way the projective connection determines an affine connection. This affine connection can be specialized to a RIEMANNian connection. We write

$$\overset{R}{\nabla}_j v^k = \partial_j v^k + \Gamma^k_{ji} v^i, \qquad \overset{R}{\nabla}_\mu v^\varkappa = \overline{A}^{\varrho\varkappa}_{\mu\sigma} \nabla_\varrho v^\sigma$$

$$\overset{R}{\nabla}_j w_i = \partial_j w_i - \Gamma^k_{ji} w_k, \qquad \overset{R}{\nabla}_\mu w_\lambda = \overline{A}^{\varrho\tau}_{\mu\lambda} \nabla_\varrho w_\tau$$

for $v^\varkappa q_\varkappa = 0$, $w_\lambda q^\lambda = 0$. In projective coordinates we can complete this connection to a projective connection by the conditions

$$\overset{R}{\nabla}_\mu q^\varkappa = 0, \qquad \overset{R}{\nabla}_\mu q_\lambda = 0, \qquad \overset{R}{P}{}^{\varkappa}_{\cdot\lambda} = 0.$$

Therefore, we have for the parameters $\overset{R}{\Pi}{}^\varkappa_{\mu\lambda}$ of this *projective* RIEMANN*ian* displacement

$$\overset{R}{\Pi}{}^\varkappa_{\mu\lambda} = \overline{A}^{ji\varkappa}_{\mu\lambda k} \Gamma^k_{ji} + \overline{A}^\varkappa_\pi \partial_\mu \overline{A}^\pi_\lambda - q^\varkappa \partial_\mu q_\lambda,$$

so that the projective connection is not symmetrical,

$$\overset{R}{S}{}_{\mu\lambda}^{\cdot\cdot\varkappa} = \partial_{[\mu} q_{\lambda]} q^\varkappa.$$

By a special assumption, as

$$P_\lambda^{\cdot\varkappa} = 4\omega q_\lambda^{\cdot\varkappa}, \qquad Q^\varkappa_{\cdot\mu} = 2\omega q_\mu^{\cdot\varkappa},$$

we can determine the complete projective connection. This assumption has been suggested by the requirements of relativity. We find under these assumptions

$$\Pi^k_{ji} = \left\{ {k \atop ji} \right\}; \qquad \Pi^k_{0i} = \omega^{-1} P^k_{\cdot i} = 4 q_i^{\cdot k},$$

$$\Pi^k_{j0} = \omega^{-1} Q^k_{\cdot j} = 2 q^k_{\cdot j}, \qquad \Pi^0_{ji} = \omega^{-1} Q_{ji} = -2 q_{ji},$$

$$\Pi^k_{00} = \Pi^0_{ji} = \Pi^0_{0i} = \Pi^0_{00} = 0.$$

In this way a RIEMANNian connection in X_n can be uniquely connected with a metrical projective connection.

6. Specialization to affine connections. Through specialization we can find several ways by which we may compare the projective dis-

[1] VAN DANTZIG: 1932 (2). — SCHOUTEN-VAN DANTZIG: 1933 (6).

placements defined, in Chapter III, by means of an A_n of given geodesic lines. For this it is first necessary to pass from the homogeneous coordinates $x^\varkappa$ to non-homogeneous ones by singling out one coordinate. This is done, as a rule, by transformations

$$\xi^\circ = \ln x^\circ, \qquad \xi^k = x^k/x^\circ, \qquad k = 1, 2, \ldots, n.$$

The transformations of $(\mathfrak{H}_{n+1})$ then pass into[1]

$$\xi^{\circ\prime} = \xi^\circ + \varphi(\xi^k), \qquad \xi^{k'} = \xi^{k'}(\xi^k),$$

so that the ξ^k are transformed like the original variables of an L_n. $\varphi(\xi^k)$ is an arbitrary function for which the value

$$(6.1) \qquad \varphi(\xi^k) = -\frac{1}{(1-c)(n+1)} \ln \Delta, \qquad \Delta = \operatorname{Det} |\partial_{k'} \xi^k|$$

has been taken[2]. This case can be considered as that of an A_n in which a scalar density φ of weight $1/(1-c)(n+1) = \mathfrak{c}$ is fixed:

$$\varphi' = \varphi \Delta^{\mathfrak{c}}.$$

In this A_n a local projective space D_n belongs to every point. It is now possible to define a number of conditions by which the $\Pi^{\varkappa}_{\mu\lambda}$ can be uniquely determined from the Γ^k_{lm} of the A_n, if the Γ^k_{lm} are given but for projective transformations. The components of the projective curvature tensor can then be identified with those of the curvature tensor $N_{\nu\lambda\mu}^{\cdot\cdot\cdot\varkappa}$. The special character of the scalar density φ permits us to work only, in this specification, with point densities of degree zero.

7. Historical remarks. The first to construct a projective displacement by associating to every point of an X_n a D_n was CARTAN[3]. It could be shown that such a displacement can be associated in a unique way with every A_n given but for projective transformations of its paths.

Another approach is due to THOMAS[4], who introduced the parameters $\Pi^{\varkappa}_{\mu\lambda}$ discussed in Ch. III, Art. 3. To these parameters belong an H_{n+1} with a limited transformation group and a covariant derivative, but not a covariant differential. This geometry corresponds to $c = -1/(n+1)$ in formula (6.1).

A third approach is due to VEBLEN[5]. He wrote the equations of the transformation in X_n as a quotient of two power series and used the members of order zero and one of these series in the definition of projective tensors. These tensors have a covariant derivative and no covariant differential. They correspond to the limiting case $c \to 1$ in (6.1).

---

[1] VEBLEN: 1933 (1).

[2] SCHOUTEN-GOŁAB: 1930 (5). — On D_n see WHITEHEAD: 1931 (39). — BORTOLOTTI: 1932 (14). — See also HLAVATÝ-GOŁAB: 1932 (6).

[3] CARTAN: 1924 (2). — SCHOUTEN: 1926 (1).

[4] THOMAS: 1926 (3). [5] VEBLEN: 1928 (3).

A fourth method was indicated by WEYL[1]; he introduced a displacement by the assumption of non-homogeneous coordinates in the local D_n.

All these methods were brought into the frame of one theory by SCHOUTEN-GOŁAB[2]. This theory involved, however, a sub-group of the group $(\mathfrak{H}_{n+1})$, and therefore still had the variable x° in a singular position. VEBLEN[3] then passed to a group holomorphic with $(\mathfrak{H}_{n+1})$ and applied it to relativity[4]. The theory was then remodelled into an independent branch of the connection theory by VAN DANTZIG[5]. SCHOUTEN and VAN DANTZIG showed how a unified field theory could be constructed on the basis of this projective connection, containing not only the gravitational and electromagnetic equations, but also the equations of SCHRÖDINGER and DIRAC[6].

Chapter VI.
Induction.

1. Ordinary surface theory. If we define, in ordinary euclidean three-space R_3, a manifold X_2 by the equations $x^\varkappa = x^\varkappa(u, v)$, $\varkappa = 1, 2, 3$, then a measurement is determined in this X_2:

$$ds^2 = E du^2 + 2F du dv + G dv^2, \qquad E = \sum_\nu \left(\frac{\partial x^\nu}{\partial u}\right)^2, \quad \text{etc.}$$

and this linear element defines a RIEMANNian connection V_2 in the X_2. We say that the RIEMANNian connection is *induced* into the X_2 by the euclidean connection of the R_3. When LEVI-CIVITA, in 1917, demonstrated the possibility of a parallel displacement in a V_2, he did it by just such a process of induction[7]. Since that time the method has often been used to obtain a differential geometry of X_m imbedded in an X_n with a certain connection. RIEMANNian geometry in a V_n leads to a RIEMANNian geometry in an imbedded X_m, a plane affine geometry in an E_n to an A_m in an imbedded X_m.[8] The "generalized absolute calculus" of VITALI is founded upon this principle[9]. We shall first show how it can be applied to an L_n.

[1] WEYL: 1929 (9). [2] SCHOUTEN-GOŁAB: 1930 (5).
[3] VEBLEN: 1929 (28) — 1933 (1). See the latter for the literature.
[4] Discussion of the theory in this state in BORTOLOTTI: 1931 (3).
[5] VAN DANTZIG: 1932 (1).
[6] SCHOUTEN and VAN DANTZIG: 1932 (3) — 1933 (6).
[7] LEVI-CIVITA: 1917 (1).
[8] SCHOUTEN: 1924 (5). — Comp. V. D. WOUDE-HAANTJES: 1933 (4). — HLAVATÝ: 1928 (9).
[9] BORTOLOTTI: 1931 (6).

2. X_m imbedded in X_n. In an X_n with original variables $\xi^\varkappa$ an X_m is *imbedded* ($m < n$). This can be done by giving n equations

$$\xi^\varkappa = \xi^\varkappa(\eta^c); \qquad \varkappa, \lambda, \mu, \nu, \cdots, = 1, \cdots, n; \qquad a, b, c, \cdots = 1, \cdots, m,$$

in which the η^c are independent coordinates and the $\xi^\varkappa$ satisfy necessary conditions as to differentiability. At a point of the X_m we have a local tangent E_n of X_n and a local tangent E_m of X_m. In the E_n we have the measuring vectors $\underset{\lambda}{e^\varkappa}$, $\overset{\varkappa}{e_\lambda}$, in the E_m $\underset{a}{e^c}$, $\overset{a}{e_c}$ and the unit tensors $A_\lambda^\varkappa$, B_a^c. The differentials $d\eta^c$ and $d\xi^\varkappa$, determining the same linear element of X_m are related by the equation

$$d\xi^\varkappa = P_c^\varkappa d\eta^c, \qquad P_c^\varkappa = \partial \xi^\varkappa / \partial \eta^c.$$

The $P_c^\varkappa$ behave as a vector in the X_n with respect to the upper index, and as a vector in the X_m with respect to the lower index:

$$P_{c'}^{\varkappa'} = A_\nu^{\varkappa'} B_{c'}^c P_c^\nu, \qquad A_\varkappa^{\varkappa'} = \partial \xi^{\varkappa'} / \partial \xi^\varkappa; \qquad B_{c'}^c = \partial \eta^c / \partial \eta^{c'}.$$

To every contravariant vector v^c of X_m is associated, in a unique way, a contravariant vector $v^\varkappa$ of X_n:

$$v^\varkappa = P_c^\varkappa v^c.$$

We may take v^c, $v^\varkappa$ as different components of the same vector $\bar{v}$ in the X_m lying in X_n.

To every covariant vector w_λ of X_n (not X_m, but X_n) is associated, in a unique way a covariant vector $'w_a$ of X_m:

$$'w_a = P_a^\mu w_\mu.$$

This vector $'w_a$ may be represented by the E_{m-1} obtained by intersecting the E_{n-1} of w_λ with the local E_m of the X_m. As

$$P_b^\varkappa B_a^b = P_a^\varkappa,$$

we may write $B_b^\varkappa$ for $P_b^\varkappa$.

We get more correspondences when the X_m is *fixed*[1] in the X_n, that is, if with every point of X_m we associate a definite local E_{n-m} of X_n at P which has no direction in common with the local E_m (in ordinary differential geometry the surface normal, in plane affine geometry the affine normal, etc.). This "pseudonormal E_{n-m}" can be defined by taking m independent covariant vectors $\overset{c}{e_\lambda}$, $c = 1, \ldots, m$ of which the E_{n-1} do not contain the local E_m of X_m. Now a quantity $Q_\lambda^c = \overset{b}{e_\lambda} \underset{b}{e^c}$ arises. We can then associate to a covariant vector w_a of X_m a covariant vector w_λ of X_n

$$w_\lambda = Q_\lambda^b w_b.$$

[1] German: "eingespannt".

The E_{n-1} of w_λ can be considered as the composition of the E_{m-1} of w_a with the pseudonormal E_{n-m}. To a contravariant vector v^ν of E_n belongs a covariant vector $'v^c$ of E_m

$$'v^c = Q^c_\mu v^\mu,$$

which can be taken as the "projection" of $v^\varkappa$ on E_m in the direction of the pseudonormal E_{n-m}. When $v^\varkappa$ lies in the E_{n-m} $'v^c = 0$, i. e. the projection is zero. As

$$Q^b_\mu B^c_b = Q^c_\mu,$$

we may write B^b_μ for Q^b_μ. The tensor B has therefore components B^a_b, B^a_λ, $B^\varkappa_c$, $B^\varkappa_\lambda$. A quantity of X_n can have components partly or wholly indicated by indices $a, b, \ldots$, when the geometrical entity it represents lies partly or wholly in the X_m.

To the E_{n-m} belongs the tensor $C^\varkappa_\lambda = A^\varkappa_\lambda - B^\varkappa_\lambda$.[1]

It is not necessary to use holonomic systems in X_n. We can even introduce a system of local E_m in X_n that need not integrate to an X_m. We can always give a definite meaning to the formulas.

3. X_m in L_n. Into the X_n we introduce a connection L_n by the displacement

$$\delta v^\varkappa = dv^\varkappa + \Gamma^\varkappa_{\mu\lambda} v^\lambda d\xi^\mu, \qquad \nabla_\mu v^\varkappa = \partial_\mu v^\varkappa + \Gamma^\varkappa_{\mu\lambda} v^\lambda.$$

If we consider $v^\varkappa$ a vector in X_m we can take the X_m-component of the contravariant vector $\delta v^\varkappa$. This component defines an *induced* displacement L_m in the X_m:

$$\overset{m}{\delta} v^c = B^c_\nu \delta v^\nu = B^c_\nu dv^\nu + B^c_\nu \Gamma^\nu_{\mu\lambda} v^\lambda d\xi^\mu$$
$$= dv^c + \Gamma^c_{ab} v^a d\eta^b.$$

In a similar manner we come to a displacement for other tensors. It is not even necessary to use holonomic systems. If we introduce into the X_n a non-holonomic system (k) (Ch. II), we can pass from

$$\delta v^k = dv^k + \Gamma^k_{ji} v^i (d\xi)^j$$

in X_n to $\overset{m}{\delta} v^c$ in X_m:

$$\overset{m}{\delta} v^c = dv^c + \Gamma^c_{ba} v^a (d\xi)^b.$$

In this case we can also consider a vector $w^\varkappa$ in the pseudonormal E_{n-m}, and define 3 other displacements[2]:

$$\overset{m}{\delta} w^r = dw^r + \Gamma^r_{bp} w^p (d\xi)^b \qquad m' = n - m,$$
$$\overset{m'}{\delta} v^c = dv^c + \Gamma^c_{qa} v^a (d\xi)^q \qquad p, q, r, \ldots = m+1, m+2, \ldots, n-m,$$
$$\overset{m'}{\delta} w^r = dw^r + \Gamma^r_{qp} w^p (d\xi)^q \qquad a, b, c, \ldots = 1, 2, \ldots, m.$$

[1] Schouten and van Kampen: 1930 (21).
[2] Weyl: 1921 (2) — Cartan: 1925 (8) p. 47.

To each belongs a covariant derivative

$$\overset{m}{\nabla}_b v^c = \partial_b v^c + \Gamma^c_{ba} v^a = B^{\mu c}_{b\nu} \nabla_\mu v^\nu ,$$

$$\overset{m}{\nabla}_b w^r = \partial_b w^r + \Gamma^r_{bp} w^p = B^\mu_b C^r_\nu \nabla_\mu w^\nu ,$$

$$\overset{m'}{\nabla}_q v^c = \partial_q v^c + \Gamma^c_{qa} v^a = C^\mu_q B^c_\nu \nabla_\mu v^\nu ,$$

$$\overset{m'}{\nabla}_q w^r = \partial_q w^r + \Gamma^r_{qp} w^p = C^{\mu r}_{q\nu} \nabla_\mu w^\nu .$$

Hence we can write these quantities in original variables of X_n:

$$\overset{m}{\nabla}_\mu v^\varkappa = B^{b\varkappa}_{\mu c} \overset{m}{\nabla}_b v^c , \qquad \overset{m}{\nabla}_\mu w^\varkappa = B^b_\mu C^\varkappa_r \overset{m}{\nabla}_b w^r ,$$

$$\overset{m'}{\nabla}_\mu v^\varkappa = C^q_\mu B^\varkappa_c \overset{m'}{\nabla}_q v^c , \qquad \overset{m'}{\nabla}_\mu w^\varkappa = C^{q\varkappa}_{\mu r} \overset{m'}{\nabla}_q w^r .$$

Such formulas also hold, as we saw, for non-holonomic systems.

From the first formula we get

$$B^\beta_\mu \nabla_\beta v^\varkappa = \overset{m}{\nabla}_\mu v^\varkappa + v^\lambda \overset{m}{H}{}^{\cdot\cdot\varkappa}_{\mu\lambda} ,$$

where

$$\overset{m}{H}{}^{\cdot\cdot\varkappa}_{\mu\lambda} = -B^{\beta\alpha}_{\mu\lambda} \nabla_\beta C^\varkappa_\alpha = B^{\beta\alpha}_{\mu\lambda} \nabla_\beta B^\varkappa_\alpha = -B^{\beta\alpha}_{\mu\lambda} \left(\nabla_\beta \overset{q}{e}_\alpha\right) \underset{q}{e}^\varkappa .$$

In similar manner

$$B^\beta_\mu \nabla_\beta w_\lambda = \overset{m}{\nabla}_\mu w_\lambda + w_\nu \overset{m}{L}{}^{\cdot\nu}_{\mu\cdot\lambda} ,$$

where

$$\overset{m}{L}{}^{\cdot\varkappa}_{\mu\cdot\lambda} = -B^{\beta\varkappa}_{\mu\gamma} \nabla_\beta C^\gamma_\lambda = B^{\beta\varkappa}_{\mu\gamma} \nabla_\beta B^\gamma_\lambda = -B^{\beta\varkappa}_{\mu\gamma} \left(\nabla_\beta \underset{q}{e}^\gamma\right) \overset{q}{e}_\lambda .$$

The tensors $\overset{m}{H}{}^{\cdot\cdot\varkappa}_{\mu\lambda}$ and $\overset{m}{L}{}^{\cdot\varkappa}_{\mu\cdot\lambda}$ are the *first* and *second* (*relative*) *curvature tensors* of the L_m in L_n. $\overset{m}{H}{}^{\cdot\cdot\varkappa}_{\mu\lambda}$ lies with its last index in E_{n-m}, with the first two indices in the L_m, $\overset{m}{L}{}^{\cdot\varkappa}_{\mu\cdot\lambda}$ lies with the middle index outside of L_m. There are also two other curvature tensors $\overset{m'}{H}{}^{\cdot\cdot\varkappa}_{\mu\lambda}$ and $\overset{m'}{L}{}^{\cdot\varkappa}_{\mu\cdot\lambda}$ belonging to the field of E_{n-m}, as $\overset{m}{H}{}^{\cdot\cdot\varkappa}_{\mu\lambda}$ and $\overset{m}{L}{}^{\cdot\varkappa}_{\mu\cdot\lambda}$ belong to the E_m.

4. *D*-notation. The differential symbols so far introduced do not exhaust all the possibilities of forming induced differentiation. It is, for instance, possible to construct a covariant differential of a tensor $v^{\cdot\varkappa}_\lambda$ lying with the first index in L_m and with the second in E_{n-m} (and therefore equivalent to $v^{\cdot p}_a$), which has these same properties. We have namely to form $d\xi^\varrho B^{\lambda\mu}_{\pi\varrho} C^\varkappa_\nu \nabla_\mu v^{\cdot\nu}_\lambda$. We can introduce a notation which takes all these possibilities into account. We define, for a $u^\varkappa$ in L_n, $v^\varkappa$ in L_m, $w^\varkappa$ in E_{n-m},

a) a differentiation with respect to L_n:

$$D_\mu p = \nabla_\mu p , \qquad D_\mu v^c = B^c_\nu \nabla_\mu v^\nu ,$$

$$D_\mu u^\varkappa = \nabla_\mu u^\varkappa , \qquad D_\mu w^r = C^r_\nu \nabla_\mu w^\nu ,$$

b) a differentiation with respect to L_m:

$$D_b p = B_b^\mu \nabla_\mu p, \qquad D_b v^c = B_{b\nu}^{\mu c} \nabla_\mu v^\nu,$$
$$D_b u^\varkappa = B_b^\mu \nabla_\mu u^\varkappa, \qquad D_b w^r = B_b^\mu C_\nu^r \nabla_\mu w^\nu,$$

c) a differentiation with respect to $L_{m'}$, the displacement defined with respect to the E_{n-m}, which as a rule is non-holonomic:

$$D_q p = C_q^\mu \nabla_\mu p, \qquad D_q v^c = C_q^\mu B_\nu^c \nabla_\mu v^\nu,$$
$$D_q u^\varkappa = C_q^\mu \nabla_\mu u^\varkappa, \qquad D_q w^r = C_{q\nu}^{\mu r} \nabla_\mu w^\nu.$$

Similar formulas can be defined with respect to covariant vectors and other quantities. The operators D satisfy the ordinary rules for differential operators with respect to addition and multiplication. Even for inner products the ordinary rules hold:

$$D_b v_a^{\cdot\varkappa} w^a = (D_b v_a^{\cdot\varkappa}) w^a + v_a^{\cdot\varkappa} (D_b w^a).$$

There should be made, however, a strict discrimination between the rules for D and the rules for ∇ so far as indices are concerned.

The relative curvature tensors can now be written

$$\overset{m}{H}{}_{ba}^{\cdot\cdot\varkappa} = D_b B_a^\varkappa = D_b D_a \xi^\varkappa, \text{[1]}$$
$$\overset{m}{L}{}_{b\cdot\lambda}^{\cdot c} = D_b B_\lambda^c.$$

5. V_m in V_n. Let a RIEMANNian manifold V_n with fundamental tensor $g_{\lambda\mu}$ be given. In an X_m in V_n a RIEMANNian connection is induced with fundamental tensor $b_{ab} = b_{ba}$

$$b_{ab} = B_{ab}^{\lambda\mu} g_{\lambda\mu}.$$

There is a tensor c_{pq} defined by $C_{pq}^{\lambda\mu} g_{\lambda\mu}$, lying in the pseudonormal E_{n-m}, here the normal R_{n-m}. We have, from RICCI's identity $\nabla_\mu g_{\lambda\nu} = 0$,

$$D_b b_{ac} = 0, \qquad D_q b_{ac} = 0, \qquad D_b c_{pr} = 0, \qquad D_q c_{pr} = 0.$$

The two curvature tensors $\overset{m}{L}{}_{\lambda\cdot\nu}^{\cdot\varkappa}$ and $\overset{m}{H}{}_{\lambda\mu}^{\cdot\cdot\varkappa}$ can be identified:

$$\overset{m}{L}{}_{ba}^{\cdot\cdot\varkappa} = D_b b_{ac} B_\lambda^c g^{\lambda\varkappa} = D_b B_a^\varkappa = \overset{m}{H}{}_{ba}^{\cdot\cdot\varkappa}.$$

We shall write $\overset{1}{L}{}_{ba}^{\cdot\cdot\varkappa}$ instead of $\overset{m}{L}{}_{ba}^{\cdot\cdot\varkappa}$, and similarly for $H_{\lambda\mu}^{\cdot\cdot\varkappa}$. If we introduce the unit vectors $\underset{q}{i}{}^\varkappa$ into the normal R_{n-m}, we have $\overset{1}{H}{}_{ba}^{\cdot\cdot\varkappa} = -B_{ba}^{\beta\alpha}\left(\nabla_\beta \overset{q}{i}_\alpha\right)\underset{q}{i}{}^\varkappa$. We also see that $\overset{1}{H}{}_{ba}^{\cdot\cdot\varkappa}$ for V_{n-1} in V_n splits into $h_{ba}\underset{n}{i}{}^\varkappa$, where $h_{ba} = h_{ab}$ is the second fundamental tensor of the V_{n-1} and $\underset{n}{i}{}^\varkappa$ the unit normal. It deserves to be mentioned that $\overset{1}{H}{}_{ba}^{\cdot\cdot\varkappa}$ is symmetrical

[1] V. D. WAERDEN: 1927 (11). — BORTOLOTTI: 1928 (6). — DUSCHEK-MAYER: 1930 (20) p. 156. — See also LAGRANGE: 1926 (5) p. 10. — Historical account in SCHOUTEN and VAN KAMPEN: 1930 (21) p. 774.

in a and b, but only when the V_{n-1} is holonomic. In the same way, if the V_m is replaced by a field of ∞^{n-m} non-holonomic fields of R_m in V_n, the $\overset{1}{H}{}_{ba}^{\cdot\cdot\varkappa}$ is not symmetrical in b and a, but it is symmetrical when the R_m can be integrated to V_m. The necessary and sufficient condition for the complete integrability of these R_m into a system of $\infty^{n-m} V_m$ is the symmetry of $\overset{1}{H}{}_{ab}^{\cdot\cdot\varkappa}$ in b and a.

The equations of Gauss and Codazzi assume the form

$$B_{dba\nu}^{\varkappa\mu\lambda c} R_{\varkappa\mu\lambda}^{\cdot\cdot\cdot\nu} = \overset{1}{R}{}_{dba}^{\cdot\cdot\cdot c} + 2\overset{1}{H}{}_{[d}^{\cdot\,c}{}_{|q|}\overset{1}{H}{}_{b]a}^{\cdot\cdot q} \quad \text{(Gauss)},$$

$$B_{dba}^{\varkappa\mu\lambda} C_\nu^r R_{\varkappa\mu\lambda}^{\cdot\cdot\cdot\nu} = -2 D_{[d} \overset{1}{H}{}_{b]a}^{\cdot\cdot r} \quad \text{(Codazzi)},$$

where $R_{\varkappa\mu\lambda}^{\cdot\cdot\cdot\nu}$ is the curvature tensor of V_n, $\overset{1}{R}{}_{dba}^{\cdot\cdot\cdot c}$ of V_m.[1]

6. Formulas of Frenet. At a point P of V_m in V_n we consider the local R_m, in which a fundamental tensor b_{ab} and a unit tensor B_a^b are defined. The connection between the local R_n and the R_m is given by the formula

$$D_b \xi^\varkappa = B_b^\varkappa .$$

The application of the operator D_b to $B_a^\varkappa$ gives the relative curvature tensor

$$D_b B_a^\varkappa = \overset{1}{H}{}_{ba}^{\cdot\cdot\varkappa} .$$

This tensor $\overset{1}{H}{}_{ba}^{\cdot\cdot\varkappa}$ lies with its first two indices in the R_m, and with its last index in an $R_{m_2} \perp R_m$. This R_{m_2} forms, together with the R_m, an $R_{m_1+m_2}$ ($m_1 = m$) in which all vectors $\partial_a \partial_b \xi^\varkappa$ lie. For the case of a V_2 in R_n this R_{m_2} is in general an R_3, in which the "curvature cone" lies, formed by the curvature vectors of all geodesics of V_2 issuing from P. If special conditions are introduced, the R_{m_2} may have fewer dimensions than three. For the case of a V_3 in R_n this R_{m_2} is at most an R_6; for a V_{m_1} we have $m_2 \leqq \frac{1}{2} m_1 (m_1 + 1)$. The R_{m_2} may be called the "*first normal space*". In the case of a curve the R_{m_2} is the principal (first) normal.

If we pass to vectors $\partial_a \partial_b \partial_c \xi^\varkappa$, we get a space $R_{m_1+m_2+m_3}$ in which all vectors lie. There is therefore in general an $R_{m_3} \perp R_{m_1+m_2}$, the *second normal space*. We denote the fundamental tensor and unit tensor of this R_{m_2} by $\overset{2}{C}_{p_2 q_2}$, $\overset{2}{B}{}_{p_2}^{r_2}$, and denote, accordingly, those of R_{m_1} by $\overset{1}{C}_{p_1 q_1}$, $\overset{1}{B}{}_{p_1}^{r_1}$. We have for $D_b \overset{2}{B}{}_{p_2}^\varkappa$

$$\overset{1}{B}{}_\nu^c D_b \overset{2}{B}{}_{p_2}^\nu = -\overset{2}{B}{}_{p_2}^\nu \overset{1}{H}{}_{b\cdot\nu}^{\cdot c} = -\overset{1}{H}{}_{b\cdot p_2}^{\cdot c};$$

$$\overset{2}{B}{}_\nu^{r_2} D_b \overset{2}{B}{}_{p_2}^\nu = \overset{2}{B}{}_{\nu\ p_2}^{r_2\lambda} D_b \overset{2}{B}{}_\lambda^\nu = -\overset{2}{B}{}_{\nu\ p_2}^{r_2\lambda} D_b (A_\lambda^\nu - \overset{2}{B}{}_\lambda^\nu) = 0 ,$$

---

[1] For A_m in A_n see Schouten: 1924 (5). — For L_m in L_n see Hlavatý: 1930 (24). — Comp. also Bortolotti: 1931 (8). — Hlavatý: 1926 (9). — Related is a paper by Ruse: 1931 (29).

so that the tensor $D_b \overset{2}{B}{}^{\varkappa}_{p_2} + \overset{1}{H}{}_{b\cdot p_2}^{\cdot\varkappa}$ lies with its ν-index in a region $\perp R_{m_1}$ and $\perp R_{m_2}$, hence (as the order of differentiation guarantees) in the R_{m_3}. We write

$$D_b \overset{2}{B}{}^{\varkappa}_{p_2} + \overset{1}{H}{}_{b\cdot p_2}^{\cdot\varkappa} = \overset{2}{H}{}_{b p_2}^{\cdot\cdot\varkappa}.$$

In this way we can continue to define with the aid of the *third, fourth* ... etc. *normal spaces*.

$$D_b \overset{l}{B}{}^{\varkappa}_{p_l} + \overset{l-1}{H}{}_{b\cdot p_l}^{\cdot\varkappa} = \overset{l}{H}{}_{b p_l}^{\cdot\cdot\varkappa},$$

as long as the left hand side does not vanish identically. The successive relative curvature tensors of higher order lie with their ν-index in the successive first, second, third, ..., normal spaces $R_{m_1}, R_{m_2}, R_{m_3}, \ldots$ The osculating l-space of every curve of the V_m, considered as a curve of V_n, lies in the $R_{m_1+m_2+\cdots+m_l}$. When for a certain m_k, $m_1+m_2+\cdots+m_k \leqq n$, the left hand side vanishes, the last equation becomes

$$D_b \overset{k}{B}{}^{\varkappa}_{p_k} = -\overset{k-1}{H}{}_{b\cdot p_k}^{\cdot\varkappa}.$$

Hence, we have, for V_m in V_n, the *formulas of* FRENET[1]

$$D_b \overset{l}{B}{}^{\varkappa}_{p_l} = -\overset{l-1}{H}{}_{b\cdot p_l}^{\cdot\varkappa} + \overset{l}{H}{}_{b p_l}^{\cdot\cdot\varkappa}, \qquad l = 1, \ldots, k$$

$$\overset{0}{H} = 0, \qquad \overset{k}{H} = 0.$$

For V_1 in R_3 these formulas are equivalent to the ordinary formulas of FRENET for space curves.

The integrability conditions of the first of these equations are the equations of GAUSS-CODAZZI. The integrability conditions of the other equations give generalizations of these equations.[2]

7. Curves in L_n. In a general L_n no orthogonality relations exist and the theory must take another form. So far, only the case of curves in L_n has been discussed. Formulas similar to those of FRENET can here be found due to the fact that, though no orthogonality relations exist, there exists on the curve an invariant parameter. Indeed, the equation

$$D w = \frac{dw}{dt} + \Gamma^{\varkappa}_{\mu\varkappa} \frac{d\xi^{\mu}}{dt} w = 0$$

determines, but for a multiplicative constant, a scalar density of weight -1 along the curve C.

---

[1] SCHOUTEN and VAN KAMPEN: 1931 (40), also 1930 (21). — Comp. DUSCHEK-MAYER: 1930 (20). — MAYER: 1931 (34). — BURSTIN: 1932 (35).

[2] For the equations of GAUSS-CODAZZI for A_m in A_n see SCHOUTEN: 1924 (5). — EISENHART: 1927 (1). — For invariants of A_m in A_n see VAN DER WAERDEN: 1927 (11). — MICHAL-BOTSFORD: 1932 (33).

We introduce, for a vector field $u^\varkappa$ defined along C, the vector

$$Du^\varkappa = \delta u^\varkappa/dt = du^\varkappa/dt + \Gamma^\varkappa_{\mu\lambda} u^\lambda d\xi^\mu/dt.$$

Then we can construct the following vectors

$$\underset{1}{v}^\varkappa = d\xi^\varkappa/dt,$$

$$\underset{2}{v}^\varkappa = D\underset{1}{v}^\varkappa, \qquad \underset{3}{v}^\varkappa = D\underset{2}{v}^\varkappa, \; \ldots, \; \underset{k}{v}^\varkappa = D\underset{k-1}{v}^\varkappa.$$

In general $k = n$, but under circumstances the series of $\underset{i}{v}^\varkappa$ may be interrupted for a $k < n$. Let us assume the general case that $k = n$, and that the vectors $\underset{1}{v}^\varkappa, \underset{2}{v}^\varkappa, \ldots, \underset{n}{v}^\varkappa$ form a linearly independent set at each point of the curve.

The parameter t will now be changed into a function $p = p(t)$ of t. Then we can define anew

$$\underset{1}{w}^\varkappa = d\xi^\varkappa/dp, \qquad \underset{2}{w}^\varkappa = D\underset{1}{w}^\varkappa, \; \ldots, \; \underset{n}{w}^\varkappa = D\underset{n-1}{w}^\varkappa.$$

The $\underset{i}{w}^\varkappa$ again form an independent set. We now determine p in such a way that

$$\underset{1}{w}^{[\varkappa_1}\underset{2}{w}^{\varkappa_2} \ldots \underset{n}{w}^{\varkappa_n]} = \left(\frac{dt}{dp}\right)^{\frac{n(n+1)}{2}} \underset{1}{v}^{[\varkappa_1}\underset{2}{v}^{\varkappa_2} \ldots \underset{n}{v}^{\varkappa_n]} =$$

$$= \left(\frac{dt}{dp}\right)^{\frac{n(n+1)}{2}} v^{12\ldots n} \underset{1}{A}^{[\varkappa_1}\underset{2}{A}^{\varkappa_2} \ldots \underset{n}{A}^{\varkappa_n]} =$$

$$= w\, \underset{1}{A}^{[\varkappa_1}\underset{2}{A}^{\varkappa_2} \ldots \underset{n}{A}^{\varkappa_n]}.$$

This is an invariant condition as $v^{1\,2\ldots n}$ is also a scalar density of weight -1. Hence

$$p = \int_{t_0}^{t} \left(\frac{v^{1\,2\ldots n}}{w}\right)^{2/n(n+1)} dt$$

is an invariant parameter along the curve, and the vectors

$$\underset{1}{w}^\varkappa = d\xi^\varkappa/dp, \qquad \underset{2}{w}^\varkappa = D\underset{1}{w}^\varkappa, \ldots, \qquad \underset{n}{w}^\varkappa = D\underset{n-1}{w}^\varkappa,$$

form an invariant set of n independent vectors in the E_n of L_n defined at a point of the curve.

As
$$D\underset{1}{w}^{[\varkappa_1}\underset{2}{w}^{\varkappa_2} \ldots \underset{n}{w}^{\varkappa_n]} = (Dw)\, \underset{1}{A}^{[\varkappa_1}\underset{2}{A}^{\varkappa_2} \ldots \underset{n}{A}^{\varkappa_n]} = 0,$$

it follows that $D\underset{n}{w}^\varkappa$ is a linear combination of $\underset{1}{w}^\varkappa, \ldots \underset{n-1}{w}^\varkappa$:

$$D\underset{n}{w}^\varkappa = \varkappa_1 \underset{1}{w}^\varkappa + \varkappa_2 \underset{2}{w}^\varkappa \ldots \varkappa_{\varkappa-1} \underset{n-1}{w}^\varkappa.$$

This equation and

$$D\underset{i}{w}^\varkappa = \underset{i+1}{w}^\varkappa, \qquad i = 1, \ldots, n-1$$

form analogues, for a curve in L_n, of the equations of FRENET. The functions $\varkappa_1, \ldots, \varkappa_{n-1}$ of p, the *affine curvatures*, can be found from the FRENET equations by determinant expressions.

These equations can be cast into a simpler form. If we introduce the n covariant vectors $\overset{i}{w}_\lambda$, $i = 1, \ldots, n$, by the relations

$$\overset{i}{w}_\lambda \underset{j}{w}^\lambda \overset{*}{=} \delta^i_j, \qquad \overset{i}{w}_\lambda \underset{i}{w}^\varkappa = A^\varkappa_\lambda$$

application of the operator D gives the equations

$$D\overset{i}{w}_\lambda = -\overset{i-1}{w}_\lambda - \varkappa_{n-i}\overset{i}{w}_\lambda, \qquad \begin{matrix} i = 1, 2, \ldots, n \\ \varkappa_0 = 0, \quad \overset{0}{w}_\lambda = 0 \end{matrix}$$

as a result of the FRENET formulas.

By further substitution

$$\overset{n}{u}_\lambda = (-1)^n \overset{n}{w}_\lambda, \quad D\overset{n}{u}_\lambda = -\overset{n-1}{u}_\lambda, \quad D\overset{n-1}{u}_\lambda = -\overset{n-2}{u}_\lambda, \ldots, D\overset{2}{u}_\lambda = -\overset{1}{u}_\lambda,$$

and transformation to the corresponding contravariant vectors $\underset{i}{u}^\varkappa$:

$$\overset{i}{u}_\lambda \underset{j}{u}^\lambda \overset{*}{=} \delta^i_j, \qquad \overset{i}{u}_\lambda \underset{i}{u}^\varkappa = A^\varkappa_\lambda,$$

we finally arrive at the formulas

$$\left.\begin{aligned} D\underset{1}{u}^\varkappa &= \underset{2}{u}^\varkappa, \\ D\underset{2}{u}^\varkappa &= -\varrho_1 \underset{1}{u}^\varkappa + \underset{2}{u}^\varkappa, \\ &\cdots\cdots\cdots \\ D\underset{k-1}{u}^\varkappa &= -\varrho_{k-2} \underset{1}{u}^\varkappa + \underset{k}{u}^\varkappa, \\ &\cdots\cdots\cdots \\ D\underset{n-1}{u}^\varkappa &= -\varrho_{n-2} \underset{1}{u}^\varkappa_2 + \underset{n}{u}^\varkappa, \\ D\underset{n}{u}^\varkappa &= -\varrho_{n-1} \underset{1}{u}^\varkappa, \end{aligned}\right\} \qquad \begin{aligned} &\text{in short:} \\ &D\underset{j}{u}^\varkappa = -\varrho_{j-1}\underset{1}{u}^\varkappa + \underset{j+1}{u}^\varkappa, \\ &j = 1, 2, \ldots, n, \\ &\varrho_0 = 0, \\ &\underset{n+1}{u}^\varkappa = 0. \end{aligned}$$

which show more outer similarity to the classical FRENET formulas. The $\varrho_1 \ldots \varrho_{n-1}$ are functions of the parameter p. The n vectors $\underset{1}{u}^\varkappa, \ldots, \underset{n}{u}^\varkappa$ form the *associate affine ennuple* at a point of the curve[1].

It should be noticed that this parameter p does not pass into the arc-length s when the L_n becomes a V_n. For this reason the $\varrho_1 \ldots \varrho_{n-1}$ do not pass directly into the ordinary curvatures of a curve in V_n.

8. P_m in P_n. In an X_n with a projective connection P_n (Ch. V) an X_m is given. Then a P_m can be introduced into this X_m by means of a system Π'^{c}_{ab}, Q'_a satisfying the homogeneity relations

$$x^d \partial_d \Pi'^{c}_{ba} = -\Pi'^{c}_{ba}, \qquad x^d \partial_d Q'_a = -Q'_a.$$

[1] HLAVATÝ: 1929 (11) — see also 1931 (10). — Extension to curves on non holonomic L_{n-1} HLAVATÝ: 1930 (24). — For curves in a WEYL connection see SCHOUTEN: 1924 (5). — HLAVATÝ: 1928 (9). — Related is a paper by WUNDHEILER: 1932 (23).

We cannot yet say that the P_m is induced into the X_m by the P_n as long as no relations are given connecting the coefficients of displacement in P_n and P_m. We can, however, already define a relative curvature tensor $H_{ba}^{\cdot\cdot\varkappa}$, a quantity of degree -1 in $x^\varkappa$,

$$H_{ba}^{\cdot\cdot\varkappa} = D_b B_a^\varkappa = \partial_b B_a^\varkappa + B_{b\,a}^{\mu\lambda} \Pi_{\mu\lambda}^\varkappa - B_c^\varkappa \Pi_{ab}'^{c},$$

satisfying the identities

$$H_{ab}^{\cdot\cdot\varkappa} x^a = P_\mu^{\cdot\varkappa} B_b^\mu - B_c^\varkappa P_b'^{\cdot c},$$

$$H_{ab}^{\cdot\cdot\varkappa} x^b = Q_{\cdot\lambda}^\varkappa B_a^\lambda - B_c^\varkappa Q_{\cdot a}'^{c}.$$

Induction begins when we relate the Q_a' and Q_λ by $Q_a' = B_a^\lambda Q_\lambda$. The $\Pi_{ba}'^{c}$ and $\Pi_{\mu\lambda}^\varkappa$ can be related by equations similar to those for L_m in L_n, if the P_m is *fixed* in P_n:

$$\Pi_{ba}'^{c} = B_{ba\varkappa}^{\mu\lambda c} \Pi_{\mu\lambda}^\varkappa + B_\varkappa^c \partial_b B_a^\varkappa.$$

Then

$$P_a'^{\cdot c} = B_{a\nu}^{\lambda c} P_\lambda^{\cdot\nu}, \qquad Q_{\cdot b}'^{c} = B_{\nu b}^{c\mu} Q_{\cdot\mu}^\nu.$$

In this case we have a second relative curvature tensor

$$L_{a\cdot\lambda}^{\cdot c} = B_{a\varkappa}^{\mu c} \nabla_\mu B_\lambda^\varkappa.$$

In such a connection there are geodesics only when certain conditions are satisfied. Indeed, a curve is here a manifold of two dimensions. Hence a geodesic must be defined by the equation $H_{ab}^{\cdot\cdot\varkappa} = 0$, or

$$\partial_b B_a^\varkappa + B_{ba}^{\mu\lambda} \Pi_{\mu\lambda}^\varkappa - B_c^\varkappa \Pi_{ab}'^{c} = 0, \qquad a, b = 0, 1$$

or

$$\frac{\partial^2 x^\varkappa}{\partial u^a \partial u^b} + \Pi_{\mu\lambda}^\varkappa \frac{\partial x^\lambda}{\partial u^a} \frac{\partial x^\mu}{\partial u^b} - \Pi_{ba}'^{c} \frac{\partial x^\varkappa}{\partial u^c} = 0$$

if for a moment we write u^a, u^b for x^a, x^b.

As totally geodesic P_2 are not in general possible, certain integrability conditions must be satisfied. We find that they are of the form

$$P_\lambda^{\cdot\varkappa} = p_\lambda x^\varkappa + (P - p_\varrho x^\varrho) A_\lambda^\varkappa, \qquad P_\lambda^{\cdot\varkappa} x^\lambda = P x^\varkappa$$

$$Q_{\cdot\mu}^\varkappa = q_\mu x^\varkappa + (P - p_\varrho x^\varrho) A_\mu^\varkappa. \qquad p_\lambda,\ q_\lambda \text{ arbitrary.}$$

These conditions are the necessary and sufficient conditions that through every point of P_n a geodesic line may pass in every direction.

In this case it can be proved that the P_n can be uniquely determined from the X_n considered as an A_n but for a projective transformation of paths[1].

[1] VAN DANTZIG: 1932 (1). — For manifolds in D_n see BORTOLOTTI: 1932 (15). — See on the paths of a projective connection also CARTAN: 1924 (2). — THOMSEN: 1930 (4). — On curves in special P_n HLAVATÝ: 1931 (9).

Notation used.

second derivative	ν	δ	d	k	s	D	$\mathfrak{D}$	$\mathfrak{d}$
first derivative	μ	γ	c	j	r	C	$\mathfrak{C}$	$\mathfrak{c}$
first covariant	λ	β	b	i	q	B	$\mathfrak{B}$	$\mathfrak{b}$
first contravariant	$\varkappa$	α	a	h	p	A	$\mathfrak{A}$	$\mathfrak{a}$

$\mathfrak{A}_n$ = group of all affine transformations in E_n with fixed origin $x^\varkappa \rightarrow x^{\varkappa'}$ (in n variables).

$\mathfrak{G}_n$ = group of all transformations $\xi^\varkappa \rightarrow \xi^{\varkappa'}$ in n variables.

$\mathfrak{H}_{n+1}$ = group of all homogeneous transformations of degree one $x^\varkappa \rightarrow x^{\varkappa'}$ in $n+1$ variables.

F = group of all point transformations $\bar{x}^\varkappa = \varrho x^\varkappa$, ϱ homogeneous, zero degree, in $x^\varkappa$.

X_n = n-dimensional regular manifold with original variables $\xi^\varkappa$ and group $\mathfrak{G}_n$ of coordinate transformations.

L_n = X_n with affine connection.

A_n = symmetrical L_n.

E_n = affine-euclidean space (a space with ordinary affine geometry).

R_n = euclidean space.

H_n = n-dimensional manifold with $n+1$ homogeneous coordinates $x^\varkappa$, $\mathfrak{H}_{n+1}$ as group of coordinate transformations, and in addition the group F of point transformations.

P_n = H_n with projective connection.

D_n = projective-euclidean P_n (a space with ordinary projective geometry).

U_n = HERMITian metrical connection.

A geometrical object is denoted by a *central letter* which is always the same, and different kinds of component-indices, e. g., $v^{\varkappa\lambda}$, $v_{\varkappa}^{\cdot\lambda}$, v_{ij}, $v_{\mathfrak{A}}^{\cdot\mathfrak{C}}$.

The word *space* is used only for manifolds with a group in the sense of KLEIN: euclidean space R_n, affine space E_n, projective space D_n.

Bibliography[1].

1827.

1. MÖBIUS, A. F.: Der baryzentrische Kalkül. Leipzig (see Ges. Werke).

1854.

1. RIEMANN, B.: Über die Tatsachen, welche der Geometrie zugrunde liegen. Ges. Werke, new edition by H. WEYL, 2. Aufl. Berlin: Julius Springer 1921.

[1] This bibliography is not complete. It lists papers referred to in the text and those that have not yet found a place in other bibliographies. Some of these bibliographies have been indicated by a star.

1884.

1. Ricci, G.: Principii di una teoria delle forme differenziale quadratiche. Ann. di Matem. (2) Vol. 12 pp. 135–167.

1889.

1. Darboux, G.: Leçons sur la théorie générale des surfaces. II. partie. 522 p. Paris: Gauthier-Villars et Fils.

1900.

1. Cotton, E.: Sur les invariants différentiels de quelques équations linéaires aux derivées partielles du second ordre. Ann. École norm. (3) Vol. 17 pp. 211–244.

1903.

1. Pascal, E.: Introduzione alla teoria delle forme differenziali di ordine qualunque. Atti Accad. naz. Lincei, Rend. (5) 12^{I}, 12^{II}, nine articles.

1913.

1. Mehmke, R.: Vorlesungen über Punkt- und Vektorenrechnung. I.: Punktrechnung. 399 S. Leipzig u. Berlin: Teubner.

1916.

1. Hessenberg, G.: Vektorielle Begründung der Differentialgeometrie. Math. Ann. Vol. 78 pp. 187–217.
2. Einstein, A.: Die Grundlage der allgemeinen Relativitätstheorie. Ann. Physik Vol. 49 pp. 769–822.

1917.

1. Levi-Civita, T.: Nozione di parallelismo in una varietà qualunque e conseguente specificazione geometrica della curvatura Riemanniana. Rend. Circ. mat. Palermo Vol. 42 pp. 173–205.

1918.

*1. Schouten, J. A.: Die direkte Analysis zur neueren Relativitätstheorie. Akad. Wetensch. Amsterdam, Verhandelingen Vol. 12 No. 6 95 p.
2. Weyl, H.: Reine Infinitesimalgeometrie. Math. Z. Vol. 2 pp. 384–411.
*3. – Raum, Zeit, Materie. 5. edition 1923. Berlin: Julius Springer.
4. Finsler, P.: Über Kurven und Flächen in allgemeinen Räumen, 121 p. Diss. Göttingen.
5. Noether, E.: Invarianten beliebiger Differentialausdrücke. Nachr. Ges. Wiss. Göttingen pp. 37–44.

1919.

1. König, R.: Über affine Geometrie. XXIV.: Ein Beitrag zu ihrer Grundlegung. Ber. sächs. Ges. d. Wiss. Leipzig Vol. 71 pp. 3–19.

1920.

1. König, R.: Beiträge zu einer allgemeinen linearen Mannigfaltigkeitslehre. Jber. Deutsch. Math.-Vereinig. Vol. 28 pp. 213–228.

1921.

1. Eddington, A. S.: A generalization of Weyl's theory of the electromagnetic and gravitational fields. Proc. Roy. Soc. A Vol. 99 pp. 104–122.
2. Weyl, H.: Zur Infinitesimalgeometrie: Einordnung der projektiven und der konformen Auffassung. Nachr. Ges. Wiss. Göttingen pp. 99–112.

3. Kaluza, Th.: Zum Unitätsproblem der Physik. S.-B. preuß. Akad. Wiss. pp. 966—972.
4. Weitzenböck, R.: Neuere Arbeiten der algebraischen Invariantentheorie. Enzykl. d. math. Wiss. Vol. III E 1.

1922.

1. Schouten, J. A.: Über die verschiedenen Arten der Übertragung, die einer Differentialgeometrie zugrunde gelegt werden können. Math. Z. Vol. 13 pp. 56—81, Nachtrag 15 p. 168.
2. Wirtinger, W.: On a general infinitesimal geometry in reference to the theory of relativity. Phil. Soc. Cambridge Trans. Vol. 22 pp. 439—448.
3. Eisenhart, L. P., and O. Veblen: The Riemann geometry and its generalization. Proc. Nat. Acad. Sci. U. S. A. Vol. 8 pp. 19—23.
4. Veblen, O.: Normal coordinates for the geometry of paths. Proc. Nat. Acad. Sci. U. S. A. Vol. 8 pp. 192—197.
5. — Projective and affine geometry of paths. Proc. Nat. Acad. Sci. U. S. A. Vol. 8 pp. 347—350 — comp. ibid. Vol. 9 pp. 3—4.
6. Cartan, E.: Sur une généralisation de la notion de courbure de Riemann et les espaces à torsion. C. R. Acad. Sci., Paris Vol. 174 pp. 593—595, also 3 other papers in Vol. 174.

1923.

1. Cartan, E.: Sur les variétés à connexion affine et la théorie de la relativité généralisée. Ann. École norm. Vol. 40 pp. 325—412; Vol. 41 (1924) pp. 1—25; Vol. 42 (1925) pp. 17—88.
2. — Les espaces à connexion conforme. Ann. Soc. Polon. math. pp. 171—221.
3. Einstein, A.: Zur allgemeinen Relativitätstheorie. S.-B. preuß. Akad. Wiss. pp. 32—38, 76—77.
4. Schouten, J. A.: Über die Einordnung der affinen Geometrie in die Theorie der höheren Übertragungen. Math. Z. Vol. 17 pp. 161—182, 183—188.
5. — Der Ricci-Kalkül. Jber. Deutsch. Math.-Vereinig. Vol. 32 pp. 91—96.
6. Veblen, O., and T. Y. Thomas: The geometry of paths. Trans. Amer. Math. Soc. Vol. 25 pp. 551—608.
7. Veblen, O.: Geometry and Physics. Science Vol. 57 pp. 129—139.
8. Weyl, H.: Mathematische Analyse des Raumproblems. Berlin: Julius Springer.
*9. Eddington, A. S.: The Mathematical Theory of Relativity. Cambridge University Press, 247 p.
10. Weitzenböck, R.: Invariantentheorie. Groningen: Noordhoff, 408 p.

1924.

1. Cartan, E.: Sur la connexion affine des surfaces. C. R. Acad. Sci., Paris Vol. 178 pp. 292—295, also 3 other papers in Vol. 178.
2. — Sur les variétés à connexion projective. Bull. Soc. Math. France Vol. 52 pp. 205—241.
3. — Les récentes généralisations de la notion d'espace. Bull. Sci. math. (2) Vol. 48 pp. 294—320.
4. — Les groupes d'holonomie des espaces généralisés. Acta math. Vol. 48 pp. 1—42.
5. Schouten, J. A.: Der Ricci-Kalkül. 311 p. Berlin: Julius Springer.
6. — Projectieve en konforme invarianten bij halfsymmetrische overbrengingen. Akad. Wetensch. Amsterdam, Versl. Vol. 34 pp. 1300—1302.

7. Friedmann, J. A., u. J. A. Schouten: Über die Geometrie der halbsymmetrischen Übertragungen. Math. Z. Vol. 21 pp. 211—223.
8. Veblen, O., and T. Y. Thomas: Extension of relative tensors. Trans. Amer. Math. Soc. Vol. 26 pp. 373—377.
9. Hlavatý, V.: Sur le déplacement linéaire du point. Věstnik. Král. Čes. Společ. Nauk. Tř II 8 p.
10. Schouten, J. A.: On the place of conformal and projective geometry in the theory of linear displacements. Akad. Wetensch. Amsterdam Proc. Vol. 27 pp. 407—424.

1925.

1. Cartan, E.: La théorie des groupes et les recherches récentes de géométrie différentielle. L'Enseignement Mathém. Vol. 24 pp. 5—18.
2. Thomas, T. Y.: On the projective and equiprojective geometries of paths. Proc. Nat. Acad. Sci. U. S. A. Vol. 11 pp. 199—203.
3. Veblen, O.: Remarks on the foundations of geometry. Bull. Amer. Math. Soc. Vol. 31 pp. 121—141.
4. Struik, D. J.: Sur quelques recherches modernes de géométrie différentielle. Rend. Semin. mat. Roma (2a) Vol. 3, 14 p.
5. Hessenberg, G.: Beispiele zur Richtungsübertragung. Jber. Deutsch. Math.-Vereinig. Vol. 33 (Angelegenheiten) pp. 93—95.
6. Thomas, T. Y.: Invariants of relative quadratic differential forms. Proc. Nat. Acad. Sci. U. S. A. Vol. 11 (1925) pp. 722—725; also Vol. 12 (1926) pp. 352 to 359.
7. Friesecke, H.: Vektorübertragung, Richtungsübertragung, Metrik. Math. Ann. Vol. 94 pp. 101—118.
8. Cartan, E.: La géométrie des espaces de Riemann. Mém. Sc. math. No. 9.

1926.

1. Schouten, J. A.: Erlanger Programm und Übertragungslehre. Rend. Circ. mat. Palermo Vol. 50 pp. 142—169.
2. — Über die Projektivkrümmung und die Konformkrümmung halbsymmetrischer Übertragungen. In mem. N. I. Lobatchevskii II. Soc. Phys. math. Kazan pp. 90—98.
3. Thomas, T. Y.: A projective theory of affinely connected manifolds. Math. Z. Vol. 25 pp. 723—733; comp. V. Hlavatý: ibid. Vol. 28 (1928) pp. 142 to 146.
4. Vranceanu, G.: Sur les espaces non-holonomes. C. R. Acad. Sci., Paris Vol. 183 pp. 852—854; also pp. 1083—1085.
5. Lagrange, R.: Calcul differéntiel absolu. Mémorial des sc. math. Paris: Gauthier-Villars.
6. Hlavatý, V.: Due úvahy metrické. Rozpravy II. Tř. Česke Akad. Ročnik 35 Čislo 22, 8 p.
7. Cartan, E.: L'axiome du plan et la Géométrie différentielle métrique. In mem. N. I. Lobačevski II. Soc. Phys. math. Kazan 9 p.
8. Veblen, O., and J. M. Thomas: Projective invariants of affine geometry of paths. Ann. of Math. Vol. 27 pp. 279—296.
9. Hlavatý, V.: Contribution au calcul différentiel absolu. Věstník král. Česke Spol. Nauk. Vol. 2 pp. 1—12.

1927.

1. Eisenhart, L. P.: Non-Riemannian geometry. Amer. Math. Soc. Colloquium Vol. VIII 184 p. New York.
2. Veblen, O.: Invariants of quadratic differential forms. London, Cambridge Math. Tracts No. 24 102 p.

*3. Struik, D. J.: On the geometry of linear displacements. Bull. Amer. Math. Soc. Vol. 33 pp. 523—564.

4. Struik, D. J., and N. Wiener: A relativistic theory of quanta. J. Math. Physics, Massachusetts Inst. Technol. Vol. 7 pp. 1—23.

5. Thomas, T. Y., and A. D. Michal: Differential invariants of relative quadratic differential forms. Ann. of Math. (2) Vol. 28 pp. 631—688.

6. Krauss, F.: Differentialinvarianten, ausgezeichnete Feldgrößen und Vektorübertragung. Math. Ann. Vol. 96 pp. 688—718.

7. Horák, Z.: Sur une généralisation de la notion de variété. Publ. Fac. Sci. Univ. Masaryk Brno. No. 86 20 p.

8. — Die Formeln für allgemeine lineare Übertragungen bei Benutzung von nichtholonomen Parametern. Nieuw Arch. Wiskde Vol. 25 pp. 193—201.

*9. Berwald, L.: Differentialinvarianten in der Geometrie. Enzykl. d. math. Wiss. Vol. III D 11 109 p.

10. Schouten, J. A.: The invariants of linear connections with different transformations. Akad. Wetensch. Amsterdam, Proc. Vol. 30 pp. 276—281.

11. van der Waerden, B.: Differentialkovarianten von n-dimensionalen Mannigfaltigkeiten in Riemannschen m-dimensionalen Räumen. Abh. math. Semin. Hamburg. Univ. Vol. 5 pp. 153—160.

12. Schouten, J. A.: Questions. Wiskund. Opgav. Amsterdam, Wisk. Gen. Vol. 14 pp. 201—207.

13. Cartan, E.: Sur les géodésiques des espaces des groupes simples. C. R. Acad. Sci., Paris Vol. 184 pp. 862—864; comp. ibid. pp. 1036—1038, 1628—1636; Vol. 185 pp. 96—98, also: Journal de Math. Vol. 6 pp. 1—119.

14. Bortolotti, E.: Sistemi assiali e connessioni nelle V_n. Atti Accad. naz. Lincei, Rend. (6) Vol. 5 pp. 390—395.

15. — Su una classe di connessioni euclidee in V_3. Rend. Circ. mat. Palermo Vol. 51 pp. 98—105.

16. Hlavatý, V.: Sur les déplacements isohodoiques. Enseign. mathem. Vol. 26 pp. 84—97.

17. — Applications des paramètres locaux. Ann. Soc. Polon. math. Vol. 5 pp. 44 to 62.

1928.

1. Douglas, J.: The general geometry of paths. Ann. of Math. Vol. 29 pp. 143 to 168.

2. Einstein, A.: Riemann-Geometrie mit Aufrechterhaltung des Begriffes des Fernparallelismus. S.-B. preuß. Akad. Wiss. pp. 217—221, also two other articles in the same periodical, 1928, 1929.

3. Veblen, O.: Projective tensors and connections. Proc. Nat. Acad. Sci. U. S. A. Vol. 14 pp. 154—166.

4. — Conformal tensors and connections. Proc. Nat. Acad. Sci. U. S. A. Vol. 14 pp. 735—745.

5. — Generalized projective geometry. J. London Math. Soc. Vol. 42 pp. 140 to 160.

6. Bortolotti, E.: Sulle varietà subordinate negli spazi a connessione affine e su di una espressione dei simboli di Riemann. Boll. Un. Mat. Ital. Vol. 7 pp. 86—94; comp. also ibid. Vol. 6 (1927) pp. 134—137.

7. Schlesinger, L.: Parallelverschiebung und Krümmungstensor. Math. Ann. Vol. 99 pp. 413—434.

8. — Parallelverschiebung und Weylsche Metrik. Jber. Deutsch. Math.-Vereinig. Vol. 37 pp. 15—18 italics.

9. Hlavatý, V.: Ein Beitrag zur Theorie der Weylschen Übertragung. Akad. Wetensch. Amsterdam, Proc. Vol. 31 pp. 878—881.

10. Hlavatý, V.: Théorie des densités dans le déplacement général. Ann. Mat. pura appl. (4) Vol. 5 pp. 73–83.

11. — Sur la déformation infinitésimale d'une courbe dans la variété métrique avec torsion. Bull. Soc. Math. France Vol. 56 pp. 18–25.

12. — Sur la seconde forme fondamentale relative aux courbes géodésiques d'une V_2^* dans V_3^*. C. R. Acad. Sci., Paris Vol. 186 pp. 1088–1090; comp. also ibid. pp. 1258–1260, 1508–1510, 1691–1694.

13. Michal, A. D.: Affinely connected function space manifolds. Amer. J. Math. Vol. 50 pp. 473–517.

14. Järnefelt, G.: Zur Affinoranalysis. Ann. Acad. Sci. Fennicae, Helsingfors (A) Vol. 28 No. 9 91 p.

15. Vranceanu, G.: Sullo scostamento geodetico nelle varietà anolonome. Atti Accad. naz. Lincei, Rend. (6) Vol. 7 pp. 134–137; comp. ibid. pp. 669 to 673.

16. Cartan, E.: Leçons sur la géométrie des espaces de Riemann. Paris: Gauthier-Villars.

17. Knebelman, M. S.: Collineations of projectively related affine manifolds. Ann. of Math. Vol. 29 pp. 389–394.

1929.

1. Schouten, J. A.: Über unitäre Geometrie. Akad. Wetensch. Amsterdam, Proc. Vol. 32 pp. 457–465.

2. Schouten, J. A., and V. Hlavatý: Zur Theorie der allgemeinen linearen Übertragung. Math. Z. Vol. 30 pp. 414–432.

3. Schouten, J. A.: Sur la signification géométrique de la propriété semi-symétrique d'une connexion intégrale qui laisse invariant le tenseur fondamental. C. R. Acad. Sci., Paris Vol. 188 pp. 955–957, 1135–1136.

4. — Über nichtholonome Übertragungen in einer L_n. Math. Z. Vol. 30, pp. 149 to 172.

5. — Zur Geometrie der kontinuierlichen Transformationsgruppen. Math. Ann. Vol. 102 pp. 244–272.

6. Schouten, J. A., u. D. van Dantzig: Über die Differentialgeometrie einer Hermiteschen Differentialform und ihre Beziehungen zu den Feldgleichungen der Physik. Akad. Wetensch. Amsterdam, Proc. Vol. 32 pp. 60–64.

7. Thomas, T. Y.: Determination of affine and metric spaces by their differential invariants. Math. Ann. Vol. 101 pp. 713–728.

*8. Bortolotti, E.: Parallelismo assoluto nelle varietà a connessioni affine e nuove vedute sulla relatività. Rend. Accad. Sci. Ist. Bologna (8) Vol. 6 (1928–1929) pp. 45–58.

9. Weyl, H.: On the foundations of general infinitesimal geometry. Bull. Amer. Math. Soc. Vol. 35 pp. 716–725.

10. Levi-Civita, T.: Vereinfachte Herstellung der Einsteinschen einheitlichen Feldgleichungen. S.-B. preuß. Akad. Wiss. pp. 137–153, english translation Blackie, London, 22 pp.

11. Hlavatý, V.: Proprietà differenziale delle curve in uno spazio a connessione lineare generale. Rend. Circ. mat. Palermo 53 p.

12. — Le parallélisme de la connexion de M. Weyl. Ann. École norm. sup. Vol. 46 pp. 73–103.

13. — Sugli invarianti differenziali di una forma bilineare miste. Ann. Mat. pura appl. (4) Vol. 6 pp. 114–126.

14. Knebelman, M. S.: Collineations and motions in generalized spaces. Amer. J. Math. Vol. 51 pp. 527–564.

15. Kawaguchi, A.: Sur les différentes connexions de l'espace fonctionnel. C. R. Acad. Sci., Paris Vol. 189 pp. 436—438.
16. Tamm, I.: Über den Zusammenhang der Einsteinschen einheitlichen Feldtheorie mit der Quantentheorie. Kon. Akad. Amsterdam Proc. Vol. 32 pp. 288—291.
17. van der Waerden, B. L.: Spinoranalyse. Nachr. Ges. Wiss. Göttingen pp. 100—109.
18. Schlesinger, L.: Über Parallelverschiebung in der Weltgeometrie. J. of Math. Vol. 161 pp. 14—20.
19. Hosokawa, T.: On the curvatures of a curve in a certain n-dimensional manifold. Sci. Rep. Tohoku Univ. Vol. 18 pp. 137—193.
20. Anderson, N. L.: An extension of Maschke's symbolism. Amer. J. Math. Vol. 51 pp. 123—138.
21. Reichenbach, H.: Zur Einordnung des neuen Einsteinschen Ansatzes über Gravitation und Elektrizität. Z. Physik Vol. 53 pp. 683—689.
22. Mandel, H.: Über den Zusammenhang zwischen der Einsteinschen Theorie des Fernparallelismus und der fünfdimensionalen Feldtheorie. Z. Physik Vol. 56 pp. 838—844.
23. Lagneau, M.: La géométrie de l'univers. J. École polytechn. (2) Vol. 27 pp. 85—157.
24. Vitali, G.: Geometria nello spazio hilbertiano. Bologna: Zanichelli.
25. Ślebodziński, W.: Note sur les variétés métriques. Prace mat. fiz. Vol. 36 (1928—1929) II pp. 61—63.
26. Zaycoff, R.: Fernparallelismus und Wellenmechanik. Z. Physik Vol. 58 pp. 833—840; Vol. 59 pp. 110—113.
27. Mandel, H.: Connection between Einstein's theory of distant parallelism and five-dimensional field theory. Z. Physik Vol. 56 pp. 11—12, 838—844.
28. Veblen, O.: Generalized projective geometry. J. London Math. Soc. Vol. 4? pp. 140—160.

1930.

1. Thomas, T. Y.: On the unified field theory. Proc. Nat. Acad. Sci. U. S. A. Vol. 16 pp. 761—776, also 5 other papers in the same per.
2. — The existence theorems in the problem of the determination of affine and metric spaces by their differential invariants. Amer. J. Math. Vol. 52 pp. 225—250.
3. — Invariantive systems of partial differential equations. Space structure as a boundary value problem. Ann of Math. (2) Vol. 31 pp. 687—713, 714—726.
4. Thomsen, G.: Topologische Fragen der Differentialgeometrie. XVI.: Über die topologischen Invarianten der Differentialgleichung $y'' = fy'^3 + gy'^2 + hy' + k$. Abh. math. Semin. Hamburg. Univ. Vol. 71 pp. 301—328.
5. Schouten, J. A., and St. Gołab: Über projektive Übertragungen und Ableitungen. Math. Z. Vol. 32 pp. 192—214. II.: Ann. Mat. pura appl. (4) Vol. 8 (1931) pp. 141—157.
6. Schouten, J. A., u. D. van Dantzig: Über unitäre Geometrie. Math. Ann. Vol. 103 pp. 319—346.
7. Graustein, W. C.: The linear element of a Riemannian V_n in terms of the Christoffel symbols of the second kind. Amer. J. Math. Vol. 52 pp. 351—356.
8. Veblen, O.: A generalization of the quadratic differential form. Quart. J. Math., Oxford Ser. Vol. 1 pp. 60—76.
*9. Cartan, E.: Notice historique sur la notion de parallélisme absolu. Math. Ann. Vol. 102 pp. 698—706.

10. Cartan, E.: Sur une problème d'équivalence et la théorie des espaces métriques généralisés. Mathematica (Cluj) Vol. 4 pp. 114–136.

11. Einstein, A.: Auf die Riemannmetrik und den Fernparallelismus gegründete einheitliche Feldtheorie. Math. Ann. Vol. 102 pp. 685–697.

12. Gołab, St.: Sopra le connessioni lineari generali. Estensione d'un teorema di Bompiani nel caso piu generale. Ann. Mat. pura appl. (4) Vol. 8 (1930–1931) pp. 283–291.

*13. — Über verallgemeinerte projektive Geometrie. Prace mat. fiz. Warszawa Vol. 37 pp. 91–153.

14. Whitehead, J. H. C.: The representation of projective spaces. Ann. of Math. Vol. 32 pp. 327–360.

15. Hlavatý, V.: Sulle coordinate geodetiche. Atti Accad. naz. Lincei, Rend. (6) Vol. 12 pp. 566–574 + art. ibid. pp. 647–654.

16. Bortolotti, E.: On parallelism and teleparallelism in curved space. J. London Math. Soc. Vol. 5 pp. 242–248.

17. Pastori, M.: Definizione intrinseca dei simboli di Christoffel e derivazione parziale dei tensori. Ist. Lombardo, Rend. Vol. 62 pp. 821–826.

18. Bargmann, V.: Über eine Verallgemeinerung des Einsteinschen Raumtyps. Z. Physik Vol. 65 pp. 830–847.

19. Einstein, A.: Zur Theorie der Räume mit Riemannmetrik und Fernparallelismus. S.-B. preuß. Akad. Wiss. pp. 401–402.

*20. Duschek, A., u. W. Mayer: Lehrbuch der Differentialgeometrie. II.: Riemannsche Geometrie. 245 p. Leipzig u. Berlin: Teubner.

21. Schouten, J. A., u. E. R. van Kampen: Zur Einbettungs- und Krümmungstheorie nichtholonomer Gebilde. Math. Ann. Vol. 103 pp. 752–783.

22. Veblen, O., and B. Hoffmann: Projective relativity. Physic. Rev. Vol. 36 pp. 810–822.

23. Eisenhart, L. P.: Projective normal coördinates. Proc. Nat. Acad. Sci. U. S. A. Vol. 16 pp. 731–740.

24. Hlavatý, V.: Sur les courbes des variétés non-holonomes. Atti Accad. naz. Lincei, Rend. (6) Vol. 12 pp. 647–654; comp. also pp. 566–574.

25. Whitehead, J. H. C., and B.V. Williams: A theorem on linear connections. Ann. of Math. (2) Vol. 31 pp. 150–157.

26. Michal, A. D.: Geodesic coordinates of order r. Bull. Amer. Math. Soc. pp. 541–546.

1931.

1. Veblen, O., and J. H. C. Whitehead: A set of axioms for differential geometry. Proc. Nat. Acad. Sci. U. S. A. Vol. 17 pp. 551–561.

2. Whitehead, J. H. C.: On linear connections. Trans. Amer. Math. Soc. Vol. 33 pp. 191–209.

3. Bortolotti, E.: Connessioni proiettive. Boll. Un. Mat. Ital. Vol. 9 (1930) pp. 288–294; Vol. 10 pp. 28–34, 83–90.

4. — Connessioni affini associate ad una $(n + 1)$-pla di congruenze in una varietà n-dimensionale. Atti Accad. naz. Lincei, Rend. (6) Vol. 14 pp. 462 to 468.

5. — Trasporti rigidi e geometria delle varietà anolonome. Boll. Un. Mat. Ital. Vol. 10 pp. 5–12.

6. — Nuova esposizione, su basi geometriche, del calcolo assoluto generalizzato del Vitali, e applicazione alle geometrie riemanniane di specie superiore. Rend. Semin. mat. Univ. Padova Vol. 2 pp. 1–48, 164–208; comp. Zbl. Math. Vol. 3 (1932) p. 322.

7. Bortolotti, E.: Una generalizzazione del calcolo assoluto rispetto a una forma differenziale quadratica specializzata. Atti Accad. naz. Lincei, Rend. (6) Vol. 13 pp. 104–108; comp. also ibid. Vol. 12 (1930) pp. 541–547.
8. — Sulle varietà subordinate. Ist. Lombardo, Rend. Vol. 2 pp. 441–463.
9. Hlavatý, V.: Projektive Invarianten einer Kurvenkongruenz und einer Kurve. Math. Z. Vol. 34 pp. 58–73.
10. — Courbes dans les espaces généralisés. Ann. Soc. Polon. math. Vol. 10 pp. 45–75.
11. Michal, A. D.: Notes on scalar extensions of tensors and properties of local coordinates. Proc. Nat. Acad. Sci. U. S. A. Vol. 17 pp. 132–136.
12. — An operation that generates absolute scalar differential invariants from tensors. Tôhoku Math. J. Vol. 34 pp. 71–77.
13. Michal, A. D., and T. S. Peterson: The invariant theory of functional forms under the groups of linear functional transformations of the third kind. Ann. of Math. (2) Vol. 32 pp. 431–450.
14. Kawaguchi, A.: Theory of connections in the generalized Finsler manifold. Proc. Imp. Acad. Jap. Vol. 7 pp. 211–214; II.: ibid. Vol. 8 (1932) pp. 340 to 343.
15. Douglas, J.: Systems of k-dimensional manifolds in an n-dimensional space. Math. Ann. Vol. 105 pp. 707–731.
16. Moisil, Gr. C.: Sur le calcul différentiel absolu des variétés plongées dans l'espace fonctionnel. Ann. Sci. Univ. Jassy Vol. 16 pp. 375–382.
17. Schouten, J. A., u. D. van Dantzig: Über unitäre Geometrien konstanter Krümmung. Akad. Wetensch. Amsterdam, Proc. Vol. 34 pp. 1293–1304.
18. Schouten, J. A.: Dirac equations in general relativity. J. Math. Physics, Massachusetts Inst. Technol. Vol. 10 pp. 239–271, 272–283.
19. Schouten, J. A., u. D. van Dantzig: Über eine vierdimensionale Deutung der neuesten Feldtheorie. Akad. Wetensch. Amsterdam, Proc. Vol. 34 pp. 1398–1407.
20. Zaycoff, R.: Über die Einsteinsche Theorie des Fernparallelismus. II. Z. Physik Vol. 67 pp. 135–137.
21. de Mira Fernandes, A.: Proprietà di alcune connessioni lineari. Atti Accad. naz. Lincei, Rend. (6) Vol. 13 pp. 179–183.
22. Einstein, A., u. W. Mayer: Systematische Untersuchung über kompatible Feldgleichungen, welche in einem Riemannschen Raume mit Fernparallelismus gesetzt werden können. S.-B. preuß. Akad. Wiss. Berlin pp. 257 to 265.
23. Podolsky, B.: A tensor form of Dirac's equation. Physic. Rev. (2) Vol. 37 pp. 1398–1405.
24. Nalli, P.: Trasporti rigidi di vettori negli spazi di Riemann. Atti Accad. naz. Lincei, Rend. (6) Vol. 13 pp. 669–675, also 2 papers ibid. pp. 734–739, 837–842.
25. Sen, R. N.: On one connexion between Levi-Civita parallelism and Einstein's teleparallelism. Proc. Edinburgh Math. Soc. (2) Vol. 2 pp. 252–255.
26. Barba, G.: Parallelismo generalizzato in una V_3. Atti Accad. naz. Lincei, Rend. (6) Vol. 14 pp. 78–81.
27. Novobatzky, K.: Erweiterung der Feldgleichungen. Z. Physik Vol. 72 pp. 683–698; comp. also ibid. Vol. 58 (1929) pp. 556–561.
28. Ruse, H. S.: An absolute partial differential calculus. Quart. J. Math., Oxford Ser. Vol. 2 pp. 190–202; comp. Zbl. Math. Vol. 2 p. 353.
29. — Normal covariant derivatives. Proc. London Math. Soc. (2) Vol. 33 pp. 66 to 76; comp. also (2) Vol. 32 pp. 87–92.

30. Cartan, E.: Géométrie euclidienne et géométrie riemannienne. Scientia pp. 393—402.

31. Laporte, O., and G. E. Uhlenbeck: Application of spinor analysis to the Maxwell and Dirac equations. Physic. Rev. Vol. 37 pp. 1380—1397; comp. ibid. pp. 1552—1554.

32. Tucker, A. W.: On generalized covariant differentiation. Ann. of Math. (2) Vol. 32 pp. 451—460.

33. Lanczos, C.: Die neue Feldtheorie Einsteins. Ergebn. d. exakt. Naturwiss. Nr. 10 pp. 97—132. Berlin: Julius Springer.

34. Mayer, W.: Beitrag zur Differentialgeometrie eindimensionaler Mannigfaltigkeiten, die in euklidischen Räumen eingebettet sind. S.-B. preuß. Akad. Wiss. p. 12.

35. Weyl, H.: Geometrie und Physik. Naturwiss. Vol. 19 pp. 49—58.

36. Whitehead, J. H. C.: The representation of projective spaces. Ann. of Math. Vol. 32 pp. 327—360.

37. Kunii, S.: On a unified theory of gravitational and eletromagnetic fields. Mem. Coll. Sci. Kyoto Univ. A Vol. 14 pp. 195—212.

38. Infeld, L.: Über eine Interpretation der neuen Einsteinschen Weltgeometrie auf dem Boden der klassischen Mechanik. Physik. Z. Vol. 32 pp. 110—112.

39. Whitehead, J. H. C.: On a class of projectively flat connections. Proc. London Math. Soc. (2) Vol. 32 pp. 93—114.

40. Schouten, J. A., u. E. R. van Kampen: Über die Krümmung einer V_m in V_n; eine Revision der Krümmungstheorie. Math. Ann. Vol. 105 pp. 144—159; comp. Zbl. Math. Vol. 2 p. 153.

1932.

*1. Dantzig, D. van: Theorie des projektiven Zusammenhangs n-dimensionaler Räume. Math. Ann. Vol. 106 pp. 400—454.

2. — Zur allgemeinen projektiven Differentialgeometrie. I.: Einordnung in die Affin-Geometrie. II.: X_{n+1} mit eingliedriger Gruppe. Akad. Wetensch. Amsterdam, Proc. Vol. 35 pp. 524—534, 535—542.

3. Schouten, J. A., u. D. van Dantzig: Zur generellen Feldtheorie. Diracsche Gleichungen und Hamiltonsche Funktion. Akad. Wetensch. Amsterdam, Proc. Vol. 35 pp. 844—853; comp. ibid. pp. 642—655.

4. Schouten, J. A., u. D. van Dantzig: Zur generellen Feldtheorie. Z. Physik Vol. 78 pp. 639—667.

5. König, R.: Zur Grundlegung der Tensorrechnung. Jber. Deutsch. Math.-Vereinig. Vol. 41 pp. 169—189.

6. Hlavatý, V., u. St. Gołab: Zur Theorie der Vektor- und Punktkonnexion. Prace mat. fiz. Warszawa Vol. 39 (1932) pp. 119—130.

*7. Hlavatý, V.: Courbes dans des espaces généralisés. Ann. Soc. Polon. math. Vol. 10 pp. 45—75.

8. Gołab, St.: Quelques problèmes métriques de la géométrie de Minkowski. Travaux Acad. des mines Cracovie F Vol. 6 79 p.

9. Thomas, T. Y.: Conformal tensors. Proc. Nat. Acad. Sci. U. S. A. Vol. 18 pp. 103—112; II.: pp. 189—193.

10. Raschewsky, R.: Eine charakteristische Eigenschaft der Schar der geodätischen Linien des zweidimensionalen affin-zusammenhängenden Raumes. Rec. math. Soc. math. Moscou Vol. 39 pp. 72—80.

11. Hosokawa, T.: Connections in the manifold admitting generalized transformations. Proc. Imp. Acad. Jap. Vol. 8 pp. 348—351.

12. Kawaguchi, A.: Die Differentialgeometrie in der verallgemeinerten Mannigfaltigkeit. Rend. Circ. mat. Palermo Vol. 56 pp. 245—276.

13. Kähler, E.: Über eine bemerkenswerte Hermitesche Metrik. Abh. math. Semin. Hamburg. Univ. Vol. 9 pp. 173—186.
14. Bortolotti, E.: Spazi proiettivamente piani. Ann. Mat. pura appl. (4) Vol. 11 pp. 111—134.
*15. — Sulle connessioni proiettive. Rend. Circ. mat. Palermo Vol. 56 pp. 1—57.
16. Dienes, P.: On the fundamental formulae of the geometry of tensor submanifolds. J. Math. pures appl. (9) Vol. 11 pp. 255—282.
*17. Veblen, O., and J. H. C. Whitehead: The foundations of differential geometry. Cambridge Tracts Vol. 29 97 p.
18. Whitehead, J. H. C.: Affine spaces of paths which are symmetric about each point. Math. Z. Vol. 35 pp. 644—659.
19. — Convex regions in the geometry of paths. Quart. J. Math., Oxford Ser. Vol. 3 pp. 33—42.
20. Vitali, G.: Sulle derivazioni covarianti. Rend. Semin. mat. Univ. Padova Vol. 3 pp. 16—27.
21. Robertson, H. P.: Groups of motions in space admitting absolute parallelism. Ann. of Math. (2) Vol. 33 pp. 496—520.
22. Cramlet, C. M.: A complete system of tensors of linear homogeneous second order differential equations. Trans. Amer. Math. Soc. Vol. 34 pp. 626—644.
23. Wundheiler, A.: Kovariante Ableitung und die Cesaroschen Unbeweglichkeitsbedingungen. Math. Z. Vol. 36 pp. 104—109.
24. Infeld, L.: Zur nichtholonomen Geometrie. Prace mat. fiz. Vol. 39 pp. 1—9.
25. Ślebodziński, W.: Sur la représentation géodésique des espaces de M. Cartan. C. R. Soc. Sci. Varsovie Vol. 24 pp. 255—261.
26. — Sur les transformations isomorphiques d'une variété à connexion affine. Prace mat. fiz. Vol. 39 pp. 55—62.
27. Straneo, P.: I tensori energetici nella teoria unitaria a geometrizzazione assoluta. Atti Accad. naz. Lincei, Rend. (6) Vol. 15 pp. 563—568.
28. de Mira Fernandes, A.: Sulla teoria unitaria dello spazio fisico. Atti Accad. naz. Lincei, Rend. (6) Vol. 15 pp. 797—804.
29. Rowe, C. H.: A characteristic property of systems of paths. Proc. Roy. Irish Acad. Vol. 40 pp. 99—106.
30. Delens, P.: Variétés à connexion affine. Généralisation de l'équation de Riccati. C. R. Acad. Sci., Paris Vol. 194 pp. 35—37.
31. Nalli, P.: Spazi di Riemann di seconda classe. Ann. Scuola norm. super. Pisa (2) Vol. 1 pp. 139—154; comp. Zbl. Math. Vol. 3 p. 171.
32. Weitzenböck, R.: Über den Reduktionssatz bei affinem und projektivem Zusammenhang. Akad. Wetensch. Amsterdam, Proc. Vol. 35 pp. 1220 to 1229.
33. Michal, A. D., and J. L. Botsford: Simultaneous differential invariants of an affine connection and a general linear connection. Proc. Nat. Acad. Sci. U. S. A. Vol. 18 pp. 558—562; comp. ibid. pp. 554—558.
34. Vranceanu, G.: Sur quelques points de la théorie des espaces non holonomes. Bul. fac. şti. Cernauţi Vol. 5 pp. 177—205.
35. Burstin, C.: Zum Einbettungsproblem. Commun. Soc. math. Kharkow (4) Vol. 5 pp. 87—95.
36. Hayden, H. A.: Sub-spaces of a space with torsion. Proc. Lond. Math. Soc. Vol. 34 pp. 27—50.

1933.

*1. Veblen, O.: Projektive Relativitätstheorie. Ergebn. d. Mathematik. Zweiter Band I 73 p. Berlin: Julius Springer.
2. Schouten, J. A.: Zur generellen Feldtheorie. Z. Physik Vol. 81 pp. 129—138, 405—417, Vol. 84 pp. 92—111.

3. THOMAS, T. Y., and E. W. TITT: Systems of partial differential equations and their characteristic surfaces. Ann. of Math. Vol. 34 pp. 1–80.
4. WOUDE, W. v. D., u. J. HAANTJES: Über das bewegte Achsensystem im affinen Raum. Akad. Wetensch. Amsterdam, Proc. Vol. 36 pp. 41–51.
5. GUGINO, E.: Sul trasporto per parallelismo lungo un circuito chiuso in uno spazio di Weyl. Atti Accad. naz. Lincei, Rend. (6) Vol. 17 pp. 45–52.
6. SCHOUTEN, J. A., and D. VAN DANTZIG: On the projective connexions and their applications to the general field-theory. Ann. of Math. (2) Vol. 34 pp. 271–312.
7. EISENHART, L. P.: Spaces admitting complete absolute parallelism. Bull. Amer. Math. Soc. Vol. 39 pp. 217–226.
*8. WEATHERBURN, C. E: The development of multidimensional differential geometry. Report Australian and N. Zealand Assoc. for Advancement of Science Vol. 21 pp. 12–28.
9. VEBLEN, O.: Geometry of two-component Spinors. Proc. Nat. Acad. Sci. U. S. A. Vol. 19, pp. 462—474.
10. VEBLEN, O.: Geometry of four-component spinors. Proc. Nat. Acad. Sci. U. S. A. Vol. 19, pp. 503—517.
11. Abhandlungen aus dem Seminar für Vektor- und Tensoranalysis. Herausgegeben von B. KAGAN. Staatsuniversität Moskau, 304 pp.